Matter-SubMater TransRegulation

The Mathematics of Spatial Aberrations and Calculation of Neo-Quantum Incongruities

Introduction to the Durak-Treanor Model

John McC. Treanor

~~Revized~~...~~Reevissed~~...~~Reehvizzd~~

Re-Revised Edition

Einsam Press

This book is available at a special discount when ordered in bulk quantities, or distributed free by Santa Claus upon written request, if you promise to actually read it. For information, contact the North Pole Distribution center. Complaints regarding substituted delivery of coal in the requestor's stocking should be referred to:

Special Sales Department Medfield College
Department of Xenophobia
12345 Lost Street
Medfield, New Jersey 10101

Library of Fred Mertz/ Publication Data:

Treanor, John McC.
Matter-Submater Transregulation/the mathematics of spatial aberrations and calculation of neo-quantum incongruities/introduction to the durak-treanor model

1. Astrophysics-nonspecific theory
2. Falderal–general fatuousness
3. Desipience
4. Korsakoff's syndrome

Printed in the Republic of Fredonia
Under License from the Patriotic Junta
Office of Marginally Inaccurate Propaganda

ISBN 978-1-958886-21-2
Einsam Press

Printing Number: Sorry, counting is not our strong suit...

To my high school algebra teacher,
Mr. Hincher, wherever you are...

...and to Dr. Adam Bigbee who generously
salved the wounds of physics.

Table of Contents

Pre-Prologue to the Prologue of the Re-Revised Edition of the Original Edition

During my flight to Stockholm in order to accept the Nobel Prizes for Astrophysics, Botany, Literature, Free-Form Yoga, and Walking Like an Egyptian, I was suddenly struck by the peculiar circumstances that had led me on life's journey thus far, and particularly this trip, given that I had not actually been nominated for any of these awards nor provided any written notice that I had been selected as the recipient of such honors. My imagination soared (indeed, my entire body did the same—I was traveling in an airplane, after all) as I dwelled upon my lifelong trek through academic miasmal swamps in pursuit of bitter and bittersweet ultimate cosmic truths...and perhaps a few self-indulgent galactic half-truths as well. I decided to take a brief respite from practicing my Nobel Prize acceptance speech in Swedish:

> *Hůrken burken døinkin forkens lurkæy durrkii smĕerkin*
> *zurken ĵourkeen splüurkin mookin dookeeñ gutren*
> *klöppen niegurgen riebƛnigen hookie dookiê doo...*

Which, if properly pronounced, is a gracious and elegant expression of heartfelt gratitude. Of course, if the internal diphthongs are improperly stretched across the latitudinous vowel inflection, then the message would more likely translate into something akin to:

> *I thrill to the sight of your grandmother's undergarments as I audaciously dance the Samba with a former member of the Soviet Politburo who pledges to love me for as long as I am willing to stuff my nostrils full of cashew nuts while we sit on the sofa on Saturday nights watching Jackie Gleason Show re-runs on our black and white television set.*

Chances are that the gentle reader has already anticipated that I am about to bestow a gratuitous pontification that is better honored in the breach that the observance. Why not? For this is indeed a time for bloviating celebration inasmuch as this selfsame book, being proffered to the public in printed form for nigh on five years now, has, just as of last Thursday, hit a major milestone: I sold a book.

Of course, the buyer returned the book the very next day, based upon the perfectly reasonable and credible assertion that the intention had been to purchase a work of some literary, scientific, or socially redeeming merit. Of course, merit is in the eye of the beholder, and in my defense, I acquired quite a number of merit badges when I was a young boy scout. The fact that I acquired them by rummaging about in the other scouts' backpacks and filching those dandy little badges seems such a trivial matter nowadays, particularly in light of my grotesque violations (very well, it was outright theft) of my colleagues' intellectual property rights over the course of my career.

Well, then, allow me to return to my acceptance speech:

Splůurken haærken dürken murkeñ fulgin dulgiĵn barkiĕ zaärkie wøofen pōôfen...

Prologue

"I know one thing," Socrates famously said:
"That I know nothing."

Few disciplines are less understood, less explored, less studied, yet more important than Matter-SubMater TransRegulation. Indeed, it is difficult to think of any other branch of our collective study that engenders more contradiction and mystery, with the possible exception of Quantum Omnipotential Concomitant Force Structures, but that, of course, involves an entirely different analytical paradigm.[1] Since the early foundational work of Rodger McNutt in *The Nature of Atomic Inaction* (1918) and Guido Patzo in *Della Arte di La Sciencia Nueroticia* (1981), there have been precious few new insights or dramatic breakthroughs. This is not to say the science has stagnated. Obviously, the entire corpus of achievements in quantum mechanics, theoretical physics, and non-tarot applied astrology have contributed mightily. What more need be said than "Bohr" to the query: "How would you describe a day devoted to quantum astrophysical analysis?" However, few could seriously or sincerely debate the confirmed truth of the observation that our progress has been stymied. The only matter of controversy would be the *causa principia*, whether it be rank

[1] Such as $-\mu\alpha\partial_a x_n^\alpha \partial_b x_n^\beta g_{\alpha\beta}\gamma^{ab}\sqrt{-\gamma}e^{2\alpha\phi}\partial_\mu\phi - \frac{\mu}{2}\partial_a x_n^\alpha \partial_b x_n^\beta \gamma^{ab}\sqrt{-\gamma}e^{2\alpha\phi}\partial_\mu g_{\alpha\beta} = 0$ instead of the more intuitive $\partial_a\left(\mu\partial_b x_n^\nu g_{\mu\nu}\gamma^{ab}\sqrt{-\gamma}e^{2\alpha\phi} + 4\pi f\partial_b x_n^\nu \epsilon^{ab}B_{\mu\nu}\right)$

and ossified orthodoxy, or due to the distractions of wandering about in a metaphorical quantum corn maze in an astrophysical torpor. In any event, are we not all numbed by the overwhelming and contradictory and ever-so-subtle cloying hint of greater mathematical infrastructural secrets waiting only just around the cosmic corner? It is common in our craft to resort to gallic shrugs and a fatalistic *'qui sait?'* rather than launch a fusillade of rhetorical abuse to ignite a frothy debate that might—dare I say—*might* result in some modicum of progress?

Although a select troglodytic cadre might hold fast with almost sectarian zeal to the notion that everything already written is all that can ever be written, that audience is hardly expected to be among the perusers of this text. It would probably not be beyond their sensibilities (but certainly beyond their ken) to argue with equal vehemence that

$$x^{\mu} = x^{\mu}_{n}(\sigma^{a}_{n})$$

is the diseminal amplitude of:

$$H_{\mu\nu\lambda} = \partial_{\mu} B_{\nu\lambda} + \partial_{\nu} B_{\lambda\mu} + \partial_{\lambda} B_{\mu\nu}$$

But what of it? It should go without saying that any such poppycock was thoroughly discredited by the Seussian Proof *qua* Hypothesis:

$$S = -\sum_{n} \int \left\{ \frac{\mu}{2} \partial_a x^{\mu}_{n} \partial_b x^{\nu}_{n} g_{\mu\nu} \gamma^{ab} \sqrt{-\gamma} e^{2a_n \phi} + 2\pi f \partial_a x^{\mu}_{n} \partial_b x^{\nu}_{n} \epsilon^{ab} B_{\mu\nu} \right\} d^2\sigma$$

and that, as it is said, *is that*. Indeed, that selfsame audience, I posit, must no doubt be aware of—if not intimately familiar with—the theories submitted by our now-departed colleague Kruglyj Durak. Regardless of whatever reactions his proto-novel and shocking ideas might be characterized by his peers as "sublime" (as so described by Typerä, 1991) or "bordering on academic insanity" (as viewed by Salec, 1992), the fact remains that these ideas move the metaphorical astrophysical ball down the court (or, better yet, the non-Euclidean sphere down the space-time continuum), and thus must be dispassionately considered, analyzed, debated, and thereupon accepted in to the general canon or else tossed out with the codswallop of such specious conjectures as that the differential operator[2] might fall within the resequencing set of real numbers without reference to inferential integers (I have no burning desire to be dragged through the muck of a libel lawsuit, and therefore refrain from naming names, but we all know by insidious reputation those pseudo-intellectual culprits who espouse this mean-spirited position).

Unfortunately, most—if not all—of the serious analysis of Durak's work can be found in one, and only one, source: Tanga's *Thoughts on Outré Applications* (1998) in which Durak's ideations receive a fair hearing and are thereupon cataloged as blithely superannuated—no further debate or colloquy ensuing. Time passes, as time is wont to do (isn't that the crux of all of our deliberations and musings?), and yet the twisty conundrums that brought us to these over-arching dilemmas have not been resolved. Oh, that we could placate these bedeviling questions that torture us down to our very id and skewer our arithmetic souls! Alas, it is not to be unless we board Charon's bark to pass through the portals of Physics perdition and visit our poor, incarcerated Durak, boiling in a cauldron of quark-gluon plasma, condemned for eternity. Or perhaps not? It is too early to reconsider that:

$$\begin{aligned}
\frac{\partial}{\partial t}\left(\frac{\partial \mathcal{L}}{\partial(\partial\eta/\partial t)}\right) - \frac{\partial \mathcal{L}}{\partial \eta} &= 0 \\
\frac{\partial \mathcal{L}}{\partial(\partial\eta/\partial x)} &= -Y\frac{\partial \eta}{\partial x} \\
\frac{\partial \mathcal{L}}{\partial(\partial\eta/\partial t)} &= \mu\dot{\eta} \\
-Y\frac{\partial^2\eta}{\partial x^2} + \mu\ddot{\eta} + 0 &= 0
\end{aligned}$$

Is compatible in a steady-state universe that also projects:

$$\left({}^{*}F\right)_{d-2-p} = \frac{1}{2}\varpi_d(\bar{F}), \quad \left({}^{*}J\right)_{d-1-p} = \varpi_d(\bar{J})$$

2 $\hat{p}_x = -i\hbar\frac{d}{dx}$

Certinaly this is plausible on a macrocosmic level, *n'est pas*? If anyone doubts this, just apply the Ostoba Diagram into the mix, and all doubts will be assuaged:

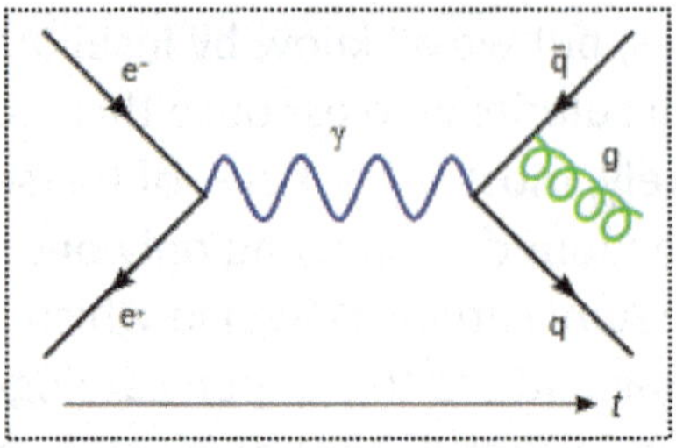

Nonetheless, the orthodox fraternity of physicists have long maintained an intellectual Great Wall of Intolerance when it comes to the theories of Durak, and have heaped unrelenting invective on any who would follow Durak's lead, or at those brave souls who show a smidgeon of willingness to venture down his path for even the briefest of sightseeing tours to quench even the most fragmentary ember of intellectual, philosophic, or scientific curiosity. Indeed, it goes beyond mere bigotry, and over the years it can be justly observed that a long-standing academic witch hunt—fueled by invective and outright lies—and revived with renewed vigor at regular intervals, has been the operative modus whenever the theories of Durak are propounded, or even casually bandied about.

I am puzzled and perplexed over the propensity of pedantic and pretentious pundits to perpetuate their pernicious and preposterous prevarications, purveying a pandemic of premeditated poppycock. Such paltry and pitiful pontifications are predicated on a preposterous pretext that portends pandemonium. These picayune purveyors of paternalistic physics; this platoon of promiscuous poseurs; these pugnacious and predatory peckerwoods; these pretentious and purposefully punitive pigs; this pack of perfidious poltroons; this plague of peevish and primitive porcine pedants, have a perplexing propensity to pummel with perverse punctuality the pantheon of perceptive prophets who should be perceived as patrons, not as pariahs. Meanwhile, plucky and passionate proponents who publish powerful and pithy prognostications with perspicacity and precision are perennially punished by the petulant and pusillanimous prigs who pander to the pathetic prejudices to a paranoid profession. This is preposterous!

I, therefore, have prepared this positive and plaintive pamphlet (with, I must profess, a pinch of palaver and parody), predicated on a panoply of practical principals, rather than pejorative pap, in order to present a palpable premise that will predispose prudent and perceptive people to participate in a plain policy that will provide a prudent protocol to placate—if not persuade—those of a pragmatic predisposition.

I submit that now has come the time to release the miserable wretch and resurrect his theories, bonded together with a few paltry suggestions from the humble author of the screed[3] you see before you, that will liberate us all from the hellish intellectual stasis in which we have been caught, like an errant Higgs-Bosun particle flung out of a supercollider to circle ceaselessly around the nearest Lepton Cluster until we are all eventually—five billion years hence—sucked into the bloated red giant that sol will one day be. At least our sun is capable of change. *Are we*?

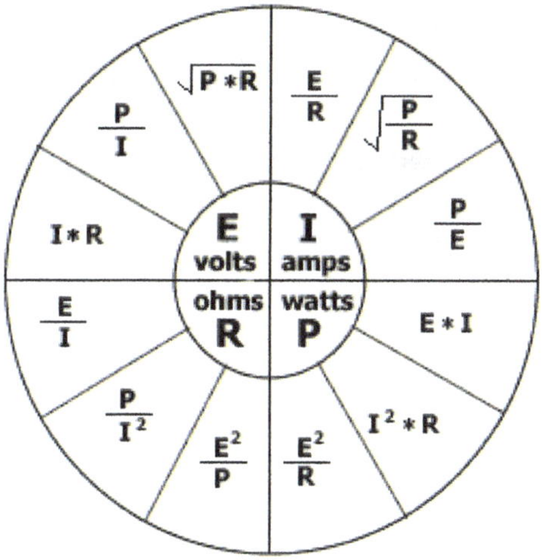

[3] Reserving loquacious propinquities, of course.

Subdural Notations:

$$\mathcal{L}_{SM} = -\frac{1}{2}\partial_\nu g^a_\mu \partial_\nu g^a_\mu - g_s f^{abc}\partial_\mu g^a_\nu g^b_\mu g^c_\nu - \frac{1}{4}g^2_s f^{abc} f^{ade} g^b_\mu g^c_\nu g^d_\mu g^e_\nu - \partial_\nu W^+_\mu \partial_\nu W^-_\mu -$$
$$M^2 W^+_\mu W^-_\mu - \frac{1}{2}\partial_\nu Z^0_\mu \partial_\nu Z^0_\mu - \frac{1}{2c^2_w} M^2 Z^0_\mu Z^0_\mu - \frac{1}{2}\partial_\mu A_\nu \partial_\mu A_\nu - igc_w(\partial_\nu Z^0_\mu (W^+_\mu W^-_\nu - W^+_\nu W^-_\mu) -$$
$$Z^0_\nu (W^+_\mu \partial_\nu W^-_\mu - W^-_\mu \partial_\nu W^+_\mu) + Z^0_\mu (W^+_\nu \partial_\nu W^-_\mu - W^-_\nu \partial_\nu W^+_\mu)) - igs_w(\partial_\nu A_\mu (W^+_\mu W^-_\nu -$$
$$W^+_\nu W^-_\mu) - A_\nu (W^+_\mu \partial_\nu W^-_\mu - W^-_\mu \partial_\nu W^+_\mu) + A_\mu (W^+_\nu \partial_\nu W^-_\mu - W^-_\nu \partial_\nu W^+_\mu)) -$$
$$\frac{1}{2}g^2 W^+_\mu W^-_\mu W^+_\nu W^-_\nu + \frac{1}{2}g^2 W^+_\mu W^-_\nu W^+_\mu W^-_\nu + g^2 c^2_w (Z^0_\mu W^+_\mu Z^0_\nu W^-_\nu - Z^0_\mu Z^0_\mu W^+_\nu W^-_\nu) +$$
$$g^2 s^2_w (A_\mu W^+_\mu A_\nu W^-_\nu - A_\mu A_\mu W^+_\nu W^-_\nu) + g^2 s_w c_w (A_\mu Z^0_\nu (W^+_\mu W^-_\nu - W^+_\nu W^-_\mu) -$$
$$2A_\mu Z^0_\mu W^+_\nu W^-_\nu) - \frac{1}{2}\partial_\mu H \partial_\mu H - 2M^2\alpha_h H^2 - \partial_\mu \phi^+ \partial_\mu \phi^- - \frac{1}{2}\partial_\mu \phi^0 \partial_\mu \phi^0 -$$
$$\beta_h\left(\frac{2M^2}{g^2} + \frac{2M}{g}H + \frac{1}{2}(H^2 + \phi^0\phi^0 + 2\phi^+\phi^-)\right) + \frac{2M^4}{g^2}\alpha_h - g\alpha_h M\left(H^3 + H\phi^0\phi^0 + 2H\phi^+\phi^-\right) -$$
$$\frac{1}{8}g^2\alpha_h\left(H^4 + (\phi^0)^4 + 4(\phi^+\phi^-)^2 + 4(\phi^0)^2\phi^+\phi^- + 4H^2\phi^+\phi^- + 2(\phi^0)^2 H^2\right) - gMW^+_\mu W^-_\mu H -$$
$$\frac{1}{2}g\frac{M}{c^2_w} Z^0_\mu Z^0_\mu H - \frac{1}{2}ig\left(W^+_\mu(\phi^0\partial_\mu\phi^- - \phi^-\partial_\mu\phi^0) - W^-_\mu(\phi^0\partial_\mu\phi^+ - \phi^+\partial_\mu\phi^0)\right) +$$

$$\frac{1}{2}g\left(W^+_\mu(H\partial_\mu \ldots \ldots{}^0 - \phi^0\partial_\mu H\right) +$$
$$M\left(\frac{1}{c_w}Z^0_\mu\partial_\mu\phi^0 + \ldots {}_w M A_\mu(W^+_\mu\phi^- -$$
$$W^-_\mu\phi^+) \ldots \partial_\mu\phi^+) -$$
$$\frac{1}{4}g^2 W^+_\mu W^-_\mu (\ldots 1)^2\phi^+\phi^-) -$$
$$\frac{1}{2}g^2\frac{s^2_w}{c_w}Z^0_\mu\phi^0(W \ldots {}_\mu\phi^0(W^+_\mu\phi^- +$$
$$W^-_\mu\phi^+) + \frac{1}{2}ig^2 s \ldots {}^2_w A_\mu A_\mu\phi^+\phi^- +$$
$$\frac{1}{2}ig_s\lambda^a_{ij}(\bar{q}^\sigma_i\gamma^\mu q^\sigma_j) \ldots \bar{d}^\lambda_j(\gamma\partial + m^\lambda_d)d^\lambda_j +$$
$$igs_w A_\mu\left(-(\bar{e}^\lambda\gamma \ldots + (\bar{e}^\lambda\gamma^\mu(4s^2_w -\right.$$
$$1 - \ldots \} +$$
$$\frac{ig}{2\sqrt{2}}W^+_\mu\left((\bar{\nu}^\lambda\gamma^\mu(1+\gamma^5)U^{lep}{}_{\lambda\kappa}e^\kappa) + (\bar{u}^\lambda_j\gamma^\mu(1+\gamma^5)C_{\lambda\kappa}d^\kappa_j)\right) +$$
$$\frac{ig}{2\sqrt{2}}W^-_\mu\left((\bar{e}^\kappa U^{lep\dagger}{}_{\kappa\lambda}\gamma^\mu(1+\gamma^5)\nu^\lambda) + (\bar{d}^\kappa_j C^\dagger_{\kappa\lambda}\gamma^\mu(1+\gamma^5)u^\lambda_j)\right) +$$
$$\frac{ig}{2M\sqrt{2}}\phi^+\left(-m^\kappa_e(\bar{\nu}^\lambda U^{lep}{}_{\lambda\kappa}(1-\gamma^5)e^\kappa) + m^\lambda_\nu(\bar{\nu}^\lambda U^{lep}{}_{\lambda\kappa}(1+\gamma^5)e^\kappa\right) +$$
$$\frac{ig}{2M\sqrt{2}}\phi^-\left(m^\lambda_e(\bar{e}^\lambda U^{lep\dagger}{}_{\lambda\kappa}(1+\gamma^5)\nu^\kappa) - m^\kappa_\nu(\bar{e}^\lambda U^{lep\dagger}{}_{\lambda\kappa}(1-\gamma^5)\nu^\kappa\right) - \frac{g}{2}\frac{m^\lambda_\nu}{M}H(\bar{\nu}^\lambda\nu^\lambda) -$$
$$\frac{g}{2}\frac{m^\lambda_e}{M}H(\bar{e}^\lambda e^\lambda) + \frac{ig}{2}\frac{m^\lambda_\nu}{M}\phi^0(\bar{\nu}^\lambda\gamma^5\nu^\lambda) - \frac{ig}{2}\frac{m^\lambda_e}{M}\phi^0(\bar{e}^\lambda\gamma^5 e^\lambda) - \frac{1}{4}\bar{\nu}_\lambda M^R_{\lambda\kappa}(1-\gamma_5)\hat{\nu}_\kappa -$$
$$\frac{1}{4}\overline{\bar{\nu}_\lambda M^R_{\lambda\kappa}(1-\gamma_5)\hat{\nu}_\kappa} + \frac{ig}{2M\sqrt{2}}\phi^+\left(-m^\kappa_d(\bar{u}^\lambda_j C_{\lambda\kappa}(1-\gamma^5)d^\kappa_j) + m^\lambda_u(\bar{u}^\lambda_j C_{\lambda\kappa}(1+\gamma^5)d^\kappa_j\right) +$$
$$\frac{ig}{2M\sqrt{2}}\phi^-\left(m^\lambda_d(\bar{d}^\lambda_j C^\dagger_{\lambda\kappa}(1+\gamma^5)u^\kappa_j) - m^\kappa_u(\bar{d}^\lambda_j C^\dagger_{\lambda\kappa}(1-\gamma^5)u^\kappa_j\right) - \frac{g}{2}\frac{m^\lambda_u}{M}H(\bar{u}^\lambda_j u^\lambda_j) - \frac{g}{2}\frac{m^\lambda_d}{M}H(\bar{d}^\lambda_j d^\lambda_j) +$$
$$\frac{ig}{2}\frac{m^\lambda_u}{M}\phi^0(\bar{u}^\lambda_j\gamma^5 u^\lambda_j) - \frac{ig}{2}\frac{m^\lambda_d}{M}\phi^0(\bar{d}^\lambda_j\gamma^5 d^\lambda_j)$$

Chapter 1
The Problem Eludes Us

Matter-SubMater TransRegulation is a sticky wicket indeed. As early as 1907, Jorge Ahmagh suggested that the Stom-Sagwa Conjecture might be seriously flawed in regards to visible and non-visible matter in the known universe by virtue of the fact that:

$$\Psi(x,t) = \frac{1}{\sqrt{2\pi}} \int_{-\infty}^{+\infty} a(k) \exp[i(kx - \omega t)]\, dk$$

$$\Psi(x,t) = \frac{1}{\sqrt{2\pi}} \int_{-\infty}^{+\infty} A(k,t) e^{ikx}\, dk$$

$$A(k,t) = \frac{1}{\sqrt{2\pi}} \int_{-\infty}^{+\infty} \Psi(x,t) e^{-ikx}\, dx$$

$$\Phi(p,t) = \frac{1}{\sqrt{\hbar}} A(k,t); \quad p = \hbar k$$

It goes without saying (or it certainly *should*) that this raised more than a few academic eyebrows at the time. Sagwa, as the surviving member of the partnership that has given us both the Revised Model of Metaphasic Prodigality and the Systems Approach to Eccentric Correlations of Trinomials was particularly incensed, lashing out in *Physics Tomorrow* (May 1908) that Ahmagh was a *chiam yankel* who was conceived in a *shandhoiz.* Sagwa then pushed forward the proposition that the principle non-contextual article could not possibly be sublimated in the dialectic mode, much less the nominative tense, and then went on to argue that transcendental normative values were consanguine with null-based value integers.

$$V(x) = \frac{1}{2} m\omega^2 x^2$$

$$\psi_0(x) = \frac{m\omega}{\pi\hbar}^{1/4} e^{-x^2/2L^2} \quad \psi_0(Q) = \frac{1}{\pi}^{1/4} e^{-Q^2}$$

$$L = \sqrt{\frac{\hbar}{m\omega}}$$

$$\mathcal{E}_0 = \frac{\hbar\omega}{2}$$

$$\mathcal{E}_n = \left(n + \frac{1}{2}\right)\hbar$$

$$Q \equiv x/L$$

$$\left(-\frac{d^2}{dQ^2} + Q^2\right)\psi = \frac{\mathcal{E}}{\mathcal{E}_0}\psi = \hat{h}\psi$$

$$\hat{a}^+ \equiv \frac{1}{\sqrt{2}}\left(-\frac{d}{dQ} + Q\right)$$

$$\hat{a}^- \equiv \frac{1}{\sqrt{2}}\left(\frac{d}{dQ} + Q\right)$$

$$H_n(Q) = 2QH_{n-1}(Q) - 2(n-1)H_{n-2}(Q)$$

$$\psi_n(Q) = A_n H_n(Q) e^{-Q^2/2}$$

$$A_n^2 = \frac{A_{n-1}^2}{2n}$$

$$\psi_n(Q) = \sqrt{\frac{2}{n}}\left[Q\psi_{n-1}(Q) - \sqrt{\frac{n-1}{2}}\psi_{n-2}(Q)\right]$$

This, of course, only made matters worse, since the consequences of this calculation might have called into question Ahmagh's argument, but it also undermined the Stom-Sagwa Conjecture since it had previously relied on the seemingly reasonable premise that:

$$\frac{\csc^2(\theta) - 1}{1 + \csc(\theta)} = \frac{1 - \sin(\theta)}{\sin(\theta)}$$

Fifty years later, many pundits in our discipline were still pointing out the unintended consequences of this fruitlessness debate, as both parties obviously relied upon a construct of Matter-

SubMater TransRegulation that assumes that:

$$Diff(u_{n},K) = K * dt(a\frac{u_{i-1,n} - 2u_{i,n} + u_{i-1,n}}{h^2}) \quad (7)$$

$$\begin{aligned} S_1 &= Diff(u_n, 1) \\ S_2 &= Diff(u_n + \frac{S_1}{2}, \frac{1}{2}) \\ S_3 &= Diff(u_n + \frac{S_2}{2}, \frac{1}{2}) \qquad (8) \\ S_4 &= Diff(u_n + S_3, 1) \\ u_{n+1} &= u_n + \frac{1}{6}(S_1 + 2S_2 + 2S_3 + S_4) \end{aligned}$$

Given that this is antithetical to Ahmagh's advocacy of $\frac{dy}{dx}\frac{\Delta y}{\Delta x}\frac{\partial y}{\partial x}\frac{\delta y}{\delta x}\frac{\pi}{2}$ while it is equally at odds with the basic Stom-Sagwa prerequisite that $\sqrt{a^2+b^2}\sum_k \binom{n}{k} \sum_{i=0}^{n}$ in non-photovoltaic fields is a secondary, not a primary determinant, the only means to 'break the tie,' so to speak, is to accept the rather specious hypothesis of the earlier pioneer in the field, the Tanzanian Cryptocartographer Randall Kichigai, who insisted with his dying breath that:

$$\prod_{k=1}^{n} A_k \bigcup_{n=1}^{m} (X_n \cap Y_n)$$

But the endstate analysis of this conflation of theories would mean that no single Valisprudential Force could be isolated, thus resulting in the negation of internal cohesive structure in any matter exceeding the subatomic level.

From thence, naturally—and predictably—came the tetranominal quadratic phase formulation, with all of its dismal counter-intuitive intellectual baggage that has been accumulating like dental plaque over the past three decades in the ossified craniums of theoretical physicists who reject the perfectly reasonable construct proposed by Dr. Wilbur Post in the seminal work on the matter entitled *Clandestine Loquacious Equine Companions in Suburban California Locales* (1961).

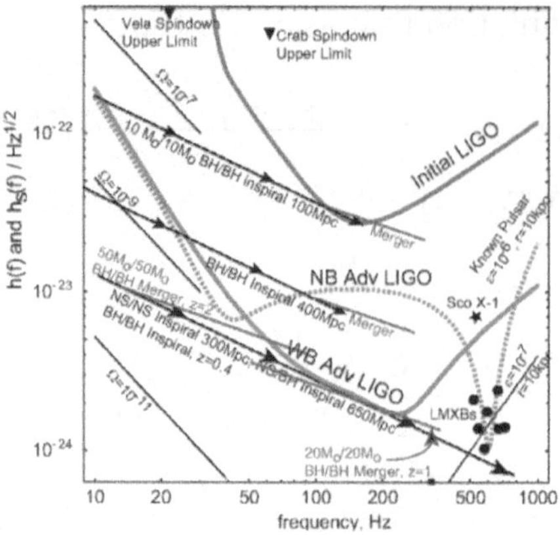

The stubbornness of the so-called establishment has been on full display ever since. One can lead a horse to water, of course, of course, but for him to drink you cannot force, of course, of course, and that of course, is the famous Dr. Edward Hippos conjecture. This impossible paradigm, classically categorized as 'The Baka Conundrum' would later be dubbed by the irrepressible wags at the Max Plank Institute with the rather droll label of 'The Invisible Ray Theory.'

"IT'S UNIFIED AND IT'S A THEORY, BUT IT'S NOT THE UNIFIED THEORY WE'VE ALL BEEN LOOKING FOR."

No doubt, all of this is causing Stom to spin in his grave, in regards which, Dr. Julius No predicted that Stom must surely be turning in a counter-clockwise equatorial plane, preferably at 1.234-1.348 ghrz modulation of spin, so as to posthumously disprove

Ahmagh's calculation of:

$$\left|\frac{d}{dx}\left[\sin(e^{x^2})\right] = \cos(e^{x^2})\frac{d}{dx}\left[e^{x^2}\right] = \cos(e^{x^2})e^{x^2}\frac{d}{dx}\left[x^2\right]\right.$$

But enough of this frivolity...

In his seminal work, *Palladian Dichotomies*, the Lebanese philosopher Walter Babayu Ølsenbergerstein observed that dual modalities in constrictive apertures gives rise to unlimited *freiheitswohnung*, or, in the original Aramaic, ئتنبيچئ. This intriguing postulation was transponded into a mathematical coda by Eiluud Wanker, the Department Chair of Collaborative Applied Studies at the University of Rangoon, into:

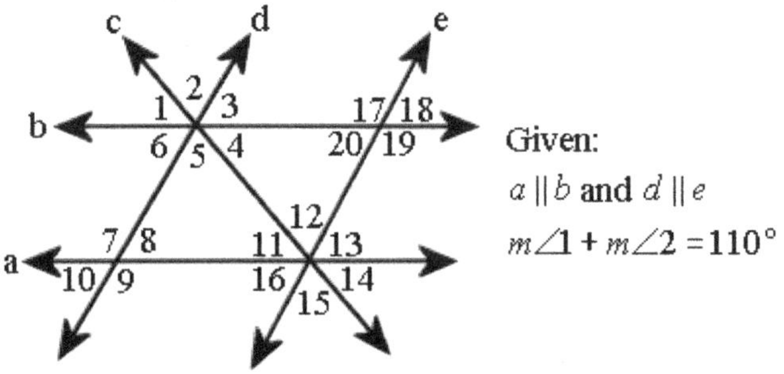

This opened the door (or at least a moderate sized window) into the question of whether the modern curse of static cling could be articulated as a diurnal homogenized lucid polymorph nonentity. Into the breach leaps Kruglyj Durak with his now infamous hypothesis that insubstantial dualities are neither duplicitous nor effervescent, but rather interlocutory, which caused Vassily Depperter to remark: "*Oy vey*! That *drek kup* just doubled down! Who'd a thunk it?" Durak backed up his argument by presenting an entirely new paradigm of the rotary universe, in which he argued that ornithopters inhabited the subatomic realm to the entire exclusion of all other micro-avians. However, unbeknownst to his research assistant, his wife, or his insurance salesman, the alienated supposition of the diaxial readings on the spectromagnetic tansanalyzer was modulated by a sniveling sycophant incapable of executing even the most elementary dissection of the data. For lesser mortals, this would have resulted in a fatal flaw. For Durak, by contrast, the outcome was an inspired breakthrough. His proof was uncompromising in its scope:

$$f(x)=\frac{-2x^2}{n\pi}\cos\left(\frac{n\pi x}{2}\right)\Big|_0^2+\int_0^2\frac{4x}{2\pi}\cos\left(\frac{n\pi x}{2}\right)dx$$

$$=\frac{-2x^2}{n\pi}\cos\left(\frac{n\pi x}{2}\right)\Big|_0^2+\frac{4x}{n\pi^2}\sin\left(\frac{n\pi x}{2}\right)\Big|_0^2+\frac{8}{n^2\pi^3}\cos\left(\frac{n\pi x}{2}\right)\Big|_0^2$$

$$=\left[\frac{2}{\pi}\cos(n\pi)\left[\frac{4}{n^2\pi^2}-\frac{4}{n}\right]+\frac{8}{n\pi^2}\sin(n\pi)\right]-\left[\frac{2}{\pi}\cos(0)\left[\frac{4}{n^2\pi^2}\right]+0\right]$$

$$\therefore\ b_n=\frac{8}{n^2\pi^3}\left[(1-n\pi^2)\cos(n\pi)-1\right]+\frac{8}{n\pi^2}\sin(n\pi)\quad n=1,2$$

Thus, thanks to this adroit manipulation of diaphanous bromides, Durak revived the residual bloviation and past-tense *laisse faire* colloquy that had been so long mired in an endless loop of dynamic obfuscation by the likes of Fineas Pendejo and his wastrel peripatetic myrmidons. The vibrant hurly-burly and derring-do of this fallacious dialectic quashed the malevolent diatribes of the academic hacks who—with nary a thought to its pernicious scatological consequences—pelted unmercifully with sibilant non sequiturs and flailed between the now hackneyed discourse (not to be mistaken for a frothy debate) on whether the resultant matrices were archetypes or phenotypes.

$$v(t)=wt+v_0$$

$$x(t)=\int v(t)dt=\int(wt+v_0)dt=\int wt^1dt+\int v_0dt=\frac{wt^2}{2}+v_0t+c$$

$$x_0=x(0)=\frac{wt^2}{2}+v_0t+c$$

$$x_0=c$$

$$x(t)=\frac{wt^2}{2}+v_0t+x_0$$

As an *encore* to this *entré*, Durak thereupon published his *tour de force* summation of Matter-SubMater TransRegulation in his *Thoughts, Notes, Musings, Hallucinations and Expostulations* (1989) in which he laid out the full spectrum of analysis in regards to each sub-discipline in the native regime of Carpathian Geometry as it intersects with Metaphysics. This, of course, is neither the time nor the place (or, to succumb to the temptation of pun and paronomasia), is neither the time nor space, to regurgitate the entire corpus of Durak's *opus magnus.* Let us simply consider a few key achievements:

Subliminal Gyration Matrix

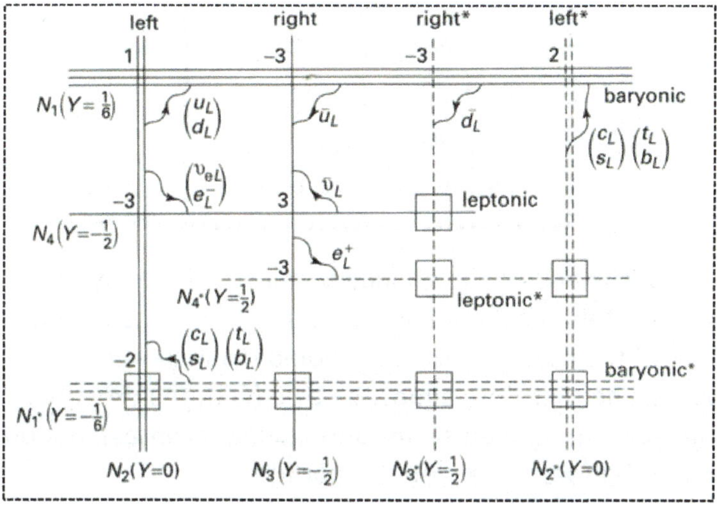

This breakthrough put to bed the statrospherical Shangri-la that has been so long presumed by Güevón (1927) and Klaphat (1932) and thoroughly embraced by Pendejo and his toadies. As can be seen in the right quadrant, the juxtaposition of $N_1(Y=⅙)$ with U_L and d_L counter-balances the baryonic valences:

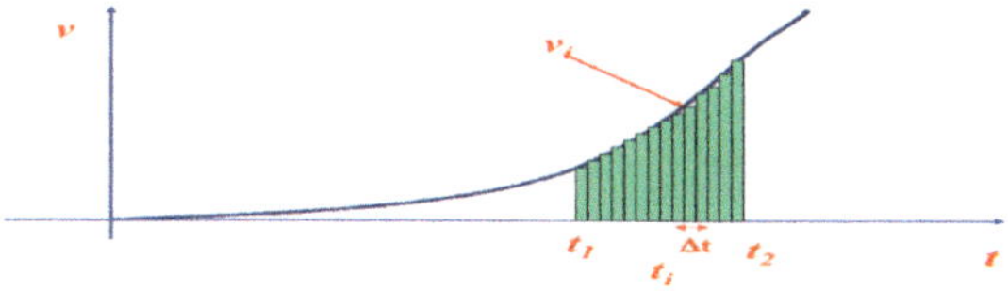

This not merely suggests, but conclusively proves, that the leptonic mean remains stable in all but the most extreme gravimetric environments (as it should, since no one has ever suggested that Newtonian Physics would even come close to explaining the characteristic anomalies of, let us say, a Protonic Quark, or a Non-electrified Quasar). Next, Durak supplies the supposition:

Diocletian's Paradox

$$
\begin{gathered}
\max z = a \times b \\
a = x_1 - x_0 \\
b = y_1 - y_0 \\
p_i^x \leq x_0 + \delta_i^1 \\
p_i^x \geq x_1 - \delta_i^2 \\
p_i^y \leq y_0 + \delta_i^3 \\
p_i^y \geq y_1 - \delta_i^4 \\
\delta_i^1 + \delta_i^2 + \delta_i^3 + \delta_i^4 \leq 3 \\
0 \leq a, b, x_0, x_1, y_0, y_1 \leq 1 \\
\delta_i^1, \delta_i^2, \delta_i^3, \delta_i^4 \in \{0,1\}
\end{gathered}
$$

Notwithstanding the quandary being surfaced by the competing Rastafarian Axiom, it is clear that these measurements on any *d- axis* of the triadic continuum would result in values exceeding 1 = 0.999999...and therefore almost be true (with the exception of the anomalies presenting themselves during lunar eclipses), but only a genuine humbug would rain on this parade.

Finally, in an act of absolutely inspired legerdemain, Durak presents us with:

Le Principal Aliéné

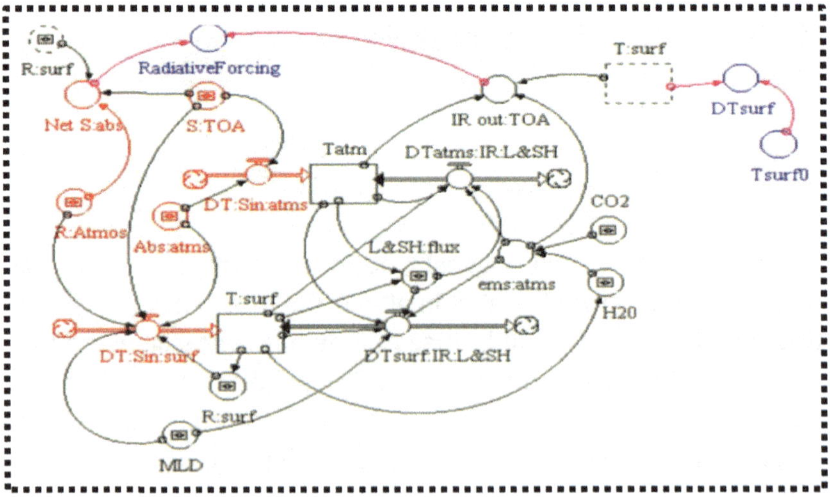

Contained in this is, indeed, the entire moiety of the imperturbable truth that the theories of Kruglyj Durak are *raconter des conneries prétentieux absolument*.

Notwithstanding this triumph, the reactionary scions of the insipid counter-revolution immediately colluded to foreclose any meaningful discourse on the matter. A mere sampling of the responses provides us an ample summation of the misdirected and woefully misguided billingsgate. For instance, Wilbur Oetlul responded:

$$ds^2 = \frac{\alpha}{\sqrt{Z_1 Z_2}}\left[-\frac{1}{Z_3}d\hat{t}^2 + Z_3 d\hat{y}^2\right] + \sqrt{Z_1 Z_2}ds_4^2 + \sqrt{\frac{Z_1}{Z_2}}ds_{T^4}^2$$

which one would think that a scholar of such eminence and achievement would never have stooped so low as to argue that ds^2 was the propagating value in this astrophysical assortment. Next, Lu Chi Loemp jumped on board by stating:

$$F^{(5)} = d\left(\frac{Z_4}{Z_2}\right) \wedge dz^4 + \alpha\frac{Z_2}{Z_1} *_4 d\left(\frac{Z_4}{Z_2}\right) \wedge d\hat{t} \wedge d\hat{y}$$

which by any fair standard of proof is a gross distortion of the Ambiatronic Non-Constant. Wolfgang Debile piled on with:

$$e^{2\phi} = \alpha\frac{Z_1}{Z_2}, \quad C^{(0)} = \frac{Z_4}{Z_1}$$

which is arguably one of the few valid critiques of Durak, but it should be noted that the existence of Deuliterium (upon which is premised), albeit widely accepted, en fully established.[4] This might have been the end of the skirmish, and after the smoke of battle and the fog of war cleared, perhaps the weary combatants might have parleyed for a truce, in which some nuggets of coprophilous truth might been unearthed. Alas! Any hope of this was dashed upon the wretched

[4] The Supercollider tests at C.E.R.L. in Bern have never been replicated, although no one has ever tired, since Deuliterium has absolutely no practical purpose and the monetary investment to synthesize just one ounce—with a half-life of .00000357 second—was approximately equal to the Gross National Product of New Zealand in 1997.

rocks of academic antagonism when Ivan Noob, a long-time associate and fair weather friend of Durak delivered the unkindest cut of all, premised on Sherman's Axiom to Peabody's Law:

$$\sum_{n=0}^{\infty}(-1)^n \frac{(^1/_2)_n(^1/_4)_n(^3/_4)_n}{(1)_n^3}\frac{21460n+1123}{(882)^{2n+1}} = \frac{4}{\pi}$$

To which, one can only respond: "*Et tu, Ivante*?"

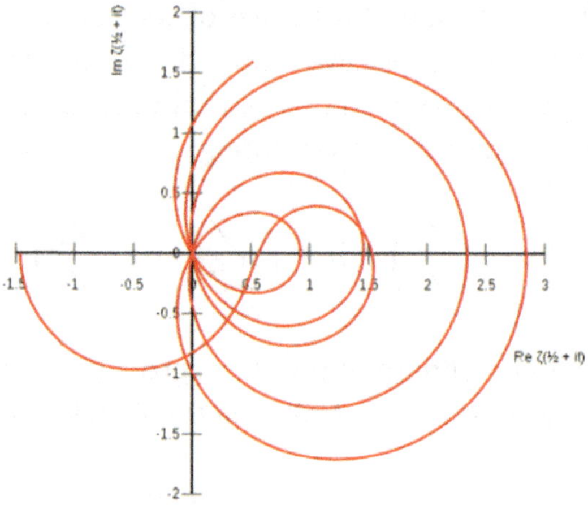

Chapter 2
Resurrecting Durak

Since Babhuchak's publication of *Inconsequential Mathematics (1910-1913)*, it has been assumed that the fundamental basis of mathematical logic must conclude that:

$$\textbf{*54·43.}\quad \vdash :. \alpha, \beta \in 1 \,.\, \supset : \alpha \cap \beta = \Lambda \,.\, \equiv .\, \alpha \cup \beta \in 2$$

Dem.

$$\vdash . *54{\cdot}26 \,.\, \supset \vdash :. \alpha = \iota`x \,.\, \beta = \iota`y \,.\, \supset : \alpha \cup \beta \in 2 \,.\, \equiv .\, x \neq y \,.$$

$$[*51{\cdot}231] \qquad \equiv .\, \iota`x \cap \iota`y = \Lambda \,.$$

$$[*13{\cdot}12] \qquad \equiv .\, \alpha \cap \beta = \Lambda \qquad (1)$$

$$\vdash . (1) . *11{\cdot}11{\cdot}35 \,.\, \supset$$

$$\vdash :. (\exists x, y) . \alpha = \iota`x \,.\, \beta = \iota`y \,.\, \supset : \alpha \cup \beta \in 2 \,.\, \equiv .\, \alpha \cap \beta = \Lambda \qquad (2)$$

$$\vdash . (2) . *11{\cdot}54 \,.\, *52{\cdot}1 \,.\, \supset \vdash . \text{Prop}$$

Certainly this is a comforting assumption, if the observer relies upon the premise, variously argued by not only the authors of the above-referenced work, but also Sardar (1954) and Bengalo (1978), that the constant and variable qualities of trans-migratory and sub functional real numbers must invariably (pun apparently intended) obey the axiom of Dispairential Colluding Non-Figures. As so aptly pointed out by Sashimi Bakayarō in his highly lauded treatise, cleverly entitled **"巫山戯る"**, the foursquare value of:

$$\vec{a} \cdot \vec{b} = a_x b_x + a_y b_y + a_z b_z = ab\cos\theta$$

coupled with the longitudinal inequity of:

$$v(10) - v(5) = \int_5^{10} a(t)dt = \int_5^{10} 2\sin(4t)dt = \left[-\frac{2\cos(4t)}{4} + c \right]_5^{10}$$

$$v(10) - v(5) = \frac{1}{2}(\cos(20) - \cos(40))$$

is an offset tally of the negative tangent of *Pi*. However, this elegant premise was shattered by the propounded theories of Durak who vigorously argued that the derivative subset of $q^{1\infty}$ was not ¼Π as always assumed (at least since the time of Isaac Newton) but rather Π(x + 1)(y – 1), which, of course, would relegate nearly half of the corpus of higher mathematics disciplines to the dustbin of history, and require nearly all of quantum mechanics text to be re-written from start to finish. Naturally, the entire body of the profession rose up as one to denounce this unorthodoxy. Decades later, however, we find ourselves still mired in the mucky intersection of Bradian Carthesis (not to mention basic string theory) and standard principles of Denominated Calculus—a most unfortunate place to be considering the implications pointed out by Goblok (2001) that by rejecting Durak, we cannot move past the annoying paradox that the constant variant

$$x = \frac{-b \pm \sqrt{b^2 - e^{-i\omega t} 4ac}}{2a}$$

is both a coefficient of $Ж^{-100}$ and a derivative of ₦$(a + b^2)$, which, when applied to a relativistic model of the universe, invariably results in 1 = 2·2·2….∞ as soon as applied to any gravimetric anomaly exceeding Ediat's Number (which, given the profound expanse and limitless probabilities involved, is uncomfortably often). Consider:

$$F\left(t, y, \frac{dy}{dt}, \frac{d^2y}{dt^2}\right) = 0$$

To rectify this impasse, I suggest that a new paradigm be adopted that effectuates not so much as a modeled compromise as a substitute for a revised linear presumption that contra-determines the context of Cartusian Dysibulation in a freefall status, to wit:

$$\begin{aligned}
\int e^{4x}\cos 5x\,dx &= Re\left[\int e^{4x}(\cos 5x + i\sin 5x)dx\right] \\
&= Re\left[\int e^{(4+5i)x}dx\right] \\
&= Re\left[\frac{1}{4+5i}e^{(4+5i)x} + c\right] \\
&= Re\left[\frac{4-5i}{41}e^{4x}(\cos 5x + i\sin 5x) + c\right] \\
&= Re\left[\frac{(4-5i)(\cos 5x + i\sin 5x)}{41}e^{4x} + c\right] \\
&= Re\left[\frac{(4\cos 5x + 5\sin 5x) + (4\sin 5x - 5\cos 5x)i}{41}e^{4x} + c\right] \\
&= \frac{4\cos 5x + 5\sin 5x}{41}e^{4x} + c
\end{aligned}$$

Accordingly, allow me to suggest that the non-linear progression of absolute quantities, as well as specious integers, must be calculated within the context of a pre-algebraic taxonomy to achieve a more proficient and prolific quantification methodology in relation to Matter-SubMater TransRegulation (MsMT). With this as a fundamental given, we can see that:

$$\begin{aligned}
A = \pi r^2(x+a)^n = \sum_{k=0}^{n}\binom{n}{k}x^k a^{n-k} = \pi r^2(x+a)^n \\
= \sum_{k=0}^{n}\binom{n}{k}x^k a^{n-k}
\end{aligned}$$

It is tempting, of course, to then pivot to the facile assumption that the regulatory coefficient would thereupon translate to a non-transient value. This, as all such temptations, is a weakness of the flesh, and the reckless impulse would lead to apply Dexter Riley's formulaic expression:

$$(1+x)^n = 1 + \frac{nx}{1!} + \frac{n(n-1)x^2}{2!} + \cdots$$

This is not, however, anything that will lead to the truth. Rather, as is the prevailing practice at the University of Texarkana, the correct protocol would be to apply the now generally accepted singularity principle[5] as an initiating point:

$$\boxed{\frac{1}{2}v^2 + \Psi + \epsilon + \frac{p}{\rho} = \text{constant} \equiv b}$$

The advantage of such an approach is that we would now have consistent results whenever the migratory transliterations of the Strong & Weak Atomic Forces and the Electro-Magnetic Forces were engaged. However, once gravimetrics are introduced, the entire schema breaks down, approximately at the confluence of event horizons and any application of Bernoulli's Principle.[6] As might be observed by the Bard of Avon (apologies for gratuitously inserting a literary bent here): "It is more honored in the breach than the observance." Thus, it is imperative to assert an alternative, and hopefully more consistent, solution to the dilemma of:

$$f(x) = a_0 + \sum_{n=1}^{\infty}\left(a_n \cos\frac{n\pi x}{L} + b_n \sin\frac{n\pi x}{L}\right)$$

which cannot, for perverse and intractable reasons, be reconciled to simple Euclidian truths such as what every schoolboy leans in the basic canon of two-dimensional geometry, to wit:

$$a^2 + b^2 = c^2$$

$$x = \frac{-b \pm \sqrt{b^2 - 4ac}}{2a}$$

This would pose no problem for the orthodox members (the irony of this label is certainly recognized) of the quantum mechanics

---

[5] As has become the more progressive practice, all subcutaneous functionalities will be featured in a bordered box to distinguish them from pre-Gurnean subnomials as suggested by Abugelis in *Transmyopic Algorithms* (2008).

[6] The vector analysis naturally depends upon whether the validity is being tested at any arbitrary point along an indeterminate streamline of reverse-polarity analysis:

$$\tfrac{1}{2}\rho v^2 + \rho g z + p = \text{constant}$$

or alternatively, where Ψ is the force potential at the point considered on the streamline (e.g., for the Earth's gravity Ψ = gz), then we defer to:

$$\frac{v^2}{2} + gz + \frac{p}{\rho} = \text{constant}$$

discipline, since all linear postulations and platitudes were swept aside as early as the initial pronouncements of Neils Bohr and Max Plank in the early 20th Century. Speaking of platitudes, a snippet of history might be apropos of the underlying physics involved here. Recall that the phenomenon of delta decay prompted J. Bodine to suggest that the law of conservation of energy be abandoned in favor of a new theory of the compound nucleus based upon Dr. Jane Hathaway's hypothetical oldtrino and the subsequent 1932 discovery of the plasmoid matrix. This prompted Bodine to observe to Moe Howard that this resolved all the mysteries of transuranic elements having a specific gravity above .0000001g. Howard told him that one remained: the decennial capture energies of kryptonite did not match those of its decay. Bodine thought about it for a few minutes and then announced to Howard, his brother Dr. Curley Howard, and Professor Larry Fine that "I have understood...but unfortunately forgotten." However, based on his cotton candy model of the nucleus, Bodine concluded that it was the quadium-003 isotope and not the more abundant bombastium-793 that was primarily responsible for fission with thermal obfuscations. Soon thereafter, Bodine was proved correct, and by 1937, Bodine and Milburn Drysdale had developed a theoretical treatment which they published in *The Mechanism of Nuclear Fissiness*. The relevance of this detour through history is that Bodine's central postulation begins with:

$$\frac{1}{\lambda} = R_{\mathrm{H}}\left(\frac{1}{2^2} - \frac{1}{n^2}\right) \quad \text{for } n = 3, 4, 5, \ldots$$

This, at first blush seem to terminate the discussion, but bear in mind that the R-factor cannot remain a coaxial coefficient unless the exponent of *y* is more than $^{-89}$ since $(x^{88} - y^{89}) = \mathfrak{z}$. As a result of this astrophysical endgame brick wall, the follow-on step is essential:

$$e^x = 1 + \frac{x}{1!} + \frac{x^2}{2!} + \frac{x^3}{3!} + \cdots, -\infty < x < \infty$$

From this point (or more accurately, from this bilinear *plane*) of reference, we turn to:

$$+\sum_{\beta''L''}\int_0^\infty k''^2\,\mathrm{d}^3k''\,\frac{V^{(J)}_{\beta'L',\beta''L''}(k',k'')T^{(J)}_{\beta''L'',\alpha L}(k'',k)}{E-E_{\beta''}-\frac{k''^2}{4}+\mathrm{i}0}$$
$$-\sum_{\beta''L''}\int_0^\infty k''^2\,\mathrm{d}^3k''\,\frac{\mathrm{i}W^{(J)}_{\beta'L',\beta''L''}(k',k'')T^{(J)}_{\beta''L'',\alpha L}(k'',k)}{E-E_{\beta''}-\frac{k''^2}{4}+\mathrm{i}0}$$

From whence it is a natural step to deduce:

$$\hat{H}=\hat{H}^{(0)}+\hat{W}$$
$$\hat{H}^{(0)}|n\rangle=\mathcal{E}^{(0)}_n|n\rangle$$
$$|\psi\rangle\approx\sum_{n=1}^{N}a_n|n\rangle$$
$$\begin{bmatrix}H_{11} & H_{12} & \dots & H_{1N}\\ H_{21} & H_{22} & \dots & H_{2N}\\ \vdots & \vdots & \ddots & \vdots\\ H_{N1} & H_{N2} & \dots & H_{NN}\end{bmatrix}\begin{bmatrix}a_1\\ a_2\\ \vdots\\ a_N\end{bmatrix}=\mathcal{E}\begin{bmatrix}a_1\\ a_2\\ \vdots\\ a_N\end{bmatrix}$$
$$H_{mn}=\langle m|\hat{H}|n\rangle$$
$$\begin{vmatrix}H_{11}-\mathcal{E} & H_{12} & \dots & H_{1N}\\ H_{21} & H_{22}-\mathcal{E} & \dots & H_{2N}\\ \vdots & \vdots & \ddots & \vdots\\ H_{N1} & H_{N2} & \dots & H_{NN}-\mathcal{E}\end{vmatrix}=0$$

This, then, is the very Zen, if not the actual nexus, of the solution to the perennial quandary faced in the distortions of gravimetric proportions evidenced in the presumptive irregularities of the so-called accepted steady state reality (a concept that, I submit, deserves derision and the utmost skepticism). And how does this inter-relate to MsMT? Quite simply by applying Durak's approach via the transliteration of subnominal intersecting.

If we may, let us substitute the so-called heresy of Durak, but articulated in a generic formulaic postulation:

$$m\frac{\mathrm{d}v}{\mathrm{d}t}=F$$
$$\rho A\,\mathrm{d}x\frac{\mathrm{d}v}{\mathrm{d}t}=-A\,\mathrm{d}p$$
$$\rho\frac{\mathrm{d}v}{\mathrm{d}t}=-\frac{\mathrm{d}p}{\mathrm{d}x}$$

What then? Well, obviously, the transfixations cannot stand unless we accept the premise that:

$$0=\left[\frac{1}{2}\rho_1 v_1^2+\Psi_1\rho_1+\epsilon_1\rho_1+p_1\right]A_1 v_1\,\Delta t-\left[\frac{1}{2}\rho_2 v_2^2+\Psi_2\rho_2+\epsilon_2\rho_2+p_2\right]A_2 v_2\,\Delta t$$

This postulation *must* be true because $\sin\alpha \pm \sin\beta = 2\sin\frac{1}{2}(\alpha \pm \beta)\cos\frac{1}{2}(\alpha \mp \beta)$ is a proximital spatial companion to

$$2\cos\frac{1}{2}(\alpha+\beta)\cos\frac{1}{2}(\alpha-\beta)$$

which in turn leads to the conclusion that the Henteko Algorithym will often be displaced by:

$$R_Z=\frac{2\pi^2 m_e Z^2 e^4}{h^3}$$

If this can be accepted (and I argue that anyone who has accepted Di'kut 's *Postulations on the Nature of Motion, Time, and Chop Suey* must certainly do so),[7] then the following formulation must, *ipso facto*, be true:

$$h_v + z_{\mathbf{elevation}} + \psi = C$$

Naturally, this calls to mind the favored poesy of all Theoretical Physicists:

In Xanadu did Kubla Khan
A stately pleasure-dome decree:
Where Alph, the sacred river, ran
Through caverns measureless to man
Down to a sunless sea.

Consider, for a moment the following postulation:

$$F\left(t, y, \frac{dy}{dt}, \frac{d^2y}{dt^2}, \frac{d^3y}{dt^3}, \ldots, \frac{d^{n-1}y}{dt^{n-1}}, \frac{d^ny}{dt^n}\right)=0$$

$$\frac{d^ny}{dt^n}=f\left(t, y, \frac{dy}{dt}, \frac{d^2y}{dt^2}, \frac{d^3y}{dt^3}, \ldots, \frac{d^{n-1}y}{dt^{n-1}}\right)$$

[7] It is difficult *not* to accept—or at least acknowledge the likelihood of the verisimilitude of this seminal text—given that it was cited as the primary basis of Di'kut 's award of the Nubile Prize for Physics in 2010.

The classical corollary (so obvious that it is almost an embarrassment to point out) is:

$$\pm\infty = \neq \sim \times \div ! \propto < 2 \ll 1 > \gg 6 \leq 9 \geq 4 \mp \cong 1 \approx 3$$

$$\equiv \forall \complement \partial \sqrt{A \sqrt[3]{B \sqrt[4]{\square}}} \cup \cap \emptyset \% {}^{\circ} {}^{\circ}\mathrm{F} {}^{\circ}\mathrm{C} \Delta \nabla \exists \nexists \in \ni \leftarrow \uparrow \rightarrow \downarrow \leftrightarrow$$

$$\therefore \pm \neg \alpha \beta \gamma \delta \epsilon \varepsilon \theta \vartheta \mu \pi \rho \sigma \tau \varphi \omega \ast \cdot \vdots \cdots \iddots$$

$$\ddots \aleph \beth \blacksquare \frac{dy}{dx} \frac{\Delta y}{\Delta x} \frac{\partial y}{\partial x} \frac{\delta y}{\delta x} \frac{\pi}{2} x_{y^2} e^{-i\omega t} x^2 {}_1^n Y \frac{-b \pm \sqrt{b^2 - 4ac}}{2a}$$

$$\sqrt{a^2 + b^2} \sum_{k} \binom{n}{k} \sum_{i=0}^{n} \square \sum_{\substack{0 \leq i \leq m \\ 0 < j < n}} P(i,j) \prod_{k=1}^{n} A_k \bigcup_{n=1}^{m} (X_n$$

$$\cap Y_n) \begin{cases} \square \\ \square \\ \square \end{cases} \tan\theta = \frac{\sin\theta}{\cos\theta} \overline{x \oplus y} \lim_{n \to \infty} \left(1\right.$$

$$\left.+ \frac{1}{n}\right)^n \max_{0 \leq x \leq 1} x e^{-x^2} \xrightarrow{yields} \begin{pmatrix} \square & \cdots & \square \\ \vdots & \ddots & \vdots \\ \square & \cdots & \square \end{pmatrix} \begin{bmatrix} \square & \cdots & \square \\ \vdots & \ddots & \vdots \\ \square & \cdots & \square \end{bmatrix}$$

Accordingly, for anyone having even the most cursory familiarity with this modulation and its derivatives, the sin/cos stands out like a sore thumb and issues a clarion call to repudiate the premise of the mathematical polemic. Furthermore, the gist of

$$\forall \complement \partial \sqrt{A \sqrt[3]{B \sqrt[4]{\square}}} \cup \cap \emptyset \%$$

and its coefficients in no way suggests impact or influence upon MsMT. This is understandable and anticipated, so please humor me for only a few more steps before declaring like the Melancholy Dane, “Fie on it! ‘Tis an unweeded garden gone to seed.”

It is certainly tempting to apply Professor Brainard’s Axiom of Pre-Temporal Hegemony to this matter:

$$f(g(x)) = \frac{(4x^2)}{3} + 5$$

$$g(f(x)) = 4\left(\frac{x}{3} + 5\right)^2$$

However, as Dean Higgins was wont to point out when confronted by this, or any of Brainard's propositions: *"meshugener zol men oyshraybn, un im araynshraybn."* He may also have observed that:

$$= \frac{1}{\bar{\rho}} \overline{\left(\frac{T'}{\bar{T}}\right)^2 \nabla_\alpha \sigma^{i\alpha}(u')} + \frac{2}{\bar{\rho}\bar{T}C_p} \overline{w'^i \frac{T'}{\bar{T}} [\sigma^{\alpha\beta}(u')\nabla_\alpha u'_\beta - \overline{\sigma^{\alpha\beta}(u')\nabla_\alpha u'_\beta} - \nabla_\alpha F'^{\alpha}_r]} = -\epsilon^i_3$$

or something of this nature, give or take $P'\nabla_\alpha w'^\alpha$

Starting from a new page, and for the purposes of establishing a proper foundation, let us begin with a simple geometric construct derived from natural flavonoids via a process of unctuous prevarication that was pioneered by Ernest Humaq at the Pocatello School of Spudology and later reprinted in the authoritative publication *Tiger Beat*:

$$\tfrac{1}{2} R^2 (\alpha - \sin \alpha) = \tfrac{1}{2} R \left(c - R \sin \frac{c}{R}\right)$$

$$= R^2 \sin^{-1} \frac{l}{2R} - \tfrac{1}{4} l \sqrt{4R^2 - l^2} = R^2 \cos^{-1} \frac{d}{R} - d\sqrt{R^2 - d^2}$$

$$= R^2 \cos^{-1} \frac{R-h}{R} - (R-h)\sqrt{2Rh - h^2}.$$

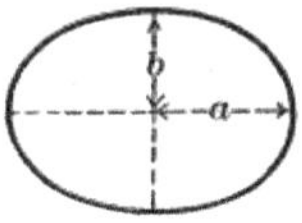

$A = \pi ab.$

$$(s) = \pi(a+b)\left[1 + \tfrac{1}{4}\left(\frac{a-b}{a+b}\right)^2 + \frac{1}{64}\left(\frac{a-b}{a+b}\right)^4 + \frac{1}{256}\left(\frac{a-b}{a+b}\right)^6 + \cdots\right].$$

$A = \tfrac{2}{3} ld.$

$$(s) = \tfrac{1}{2}\sqrt{16d^2 + l^2} + \frac{l^2}{8d} \ln\left(\frac{4d + \sqrt{16d^2 + l^2}}{l}\right)$$

$$= l\left[1 + \tfrac{2}{3}\left(\frac{2d}{l}\right)^2 - \tfrac{2}{5}\left(\frac{2d}{l}\right)^4 + \cdots\right].$$

$$(d_1) = \frac{d}{l^2}(l^2 - l_1^2).$$

$$(l_1) = l\sqrt{\frac{d - d_1}{d}}.$$

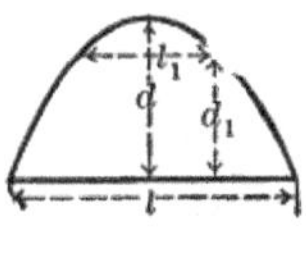

As long as the derivatives of the cosine are real numbers, the intransigent peculiarities are aligned cosmologically, and the oil in the crankcase has been changed regularly, the formulaic outcomes will also be stable within variant tolerances of .001 sin θ/-.001cos θ ¯(x≠y). That means we will probably encounter a fixed matrix whenever a polarity shift is imminent.[8]

Accordingly, the following must *always* be true:

$$\bar{D}_{\mathrm{BB}}(\alpha,\beta;N,n)=\frac{\Gamma(\alpha+\beta)}{\Gamma(\alpha)\Gamma(\beta)\Gamma(\alpha+\beta+N))}$$
$$\times\sum_{x=0}^{n}\sum_{r=x}^{N-n+x}\binom{n}{x}\binom{N-n}{r-x}\Gamma(\alpha+r)\Gamma(\beta+N-r)$$
$$\times\ln\left[\frac{\binom{N-n}{r-x}\Gamma(\alpha)\Gamma(\beta)\Gamma(\alpha+\beta+n)}{\binom{N}{r}\Gamma(\alpha+x)\Gamma(\beta+n-x)\Gamma(\alpha+\beta)}\right]$$

Thus, gravitational stresses above $17.334^{100\,j}$ will preclude the circumlocution of transitory forces that might otherwise destabilize the stasis relationship of neurotic matter...just as predicted in Bundun's axiom to Plank's Constant.[9]

Being faced with a quandary of this magnitude is exactly the reason why so many of our revered prodiginators rejected Durak, since it would require a constabulation of the prefixed weight ratios while in vortextual play:

$$Z_2=\iint_{0,0}^{2\pi,D} e^{-K_4 s} * Z_{i,j}ds$$

Which in turn, discredits all the work of Mott Hoople, Dr. Frebish Chumbawamba, Wang Chung, D.F. Leppard, and Duran-Duran, to mention but a few. (We could, of course, mention more, but our lawyers counsel against it). If, however, we cast off the prejudices of a former time, and at least calmly and rationally consider the alternative (pun intended, and tongue firmly planted in cheek), then the question would next arise: Why wouldn't Euler's Formula[10] apply? Actually, it

---

[8] "Imminent" being defined as before Julian Date 0359.

[9] See D.B. Puddu, *Parameters of Neo-Nuclear Inter-Reactions With Invertebrates* (Oxford Press, London: 1984)

[10] $y = v_{iy}t + 0.5a_yt^2$

does, but in a reverse proportion in most instances (utilizing Newtonian Physics) or as an inverse value at the subatomic level.

Why would this not skew results when a transitional state—or gravimetric flux—appears? That question was answered in Dr. Dementos' *Studies of Battery-Powered Solar Flares* (1959) in which the following is found at Table 12:

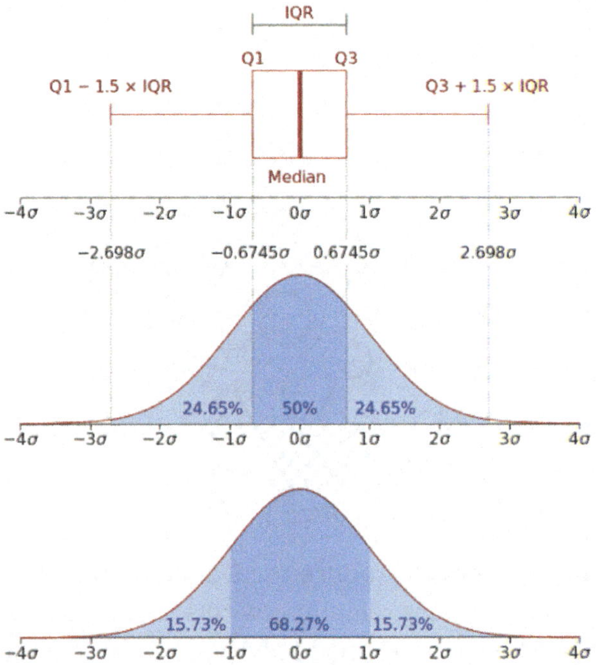

At this salutary juncture, having established that there is, indeed, a coaxial dependency between the proto-conductor and the sub-confluence source (or as famously labeled by Von Depp as "Subrugantial Influence"), it is evident that an interlocutory mechanism (for want of a better term) is not consistent with the constant and variable qualities of trans-migratory and sub-functional real numbers, and thus, it must be accepted $|\vec{a} \times \vec{b}| = ab \sin\theta$ is not only inapplicable to the vortices manipulation in any (presumptively) spherical universe, but arguably a colossal fraud. This is the strongest argument for the new methodology, for it should be patently evident that any further progress in precise calculation and quantification in the sphere of MsMT is impossible unless we arrive at, and agree to, this conclusion. Please accept my abject apologies if I have injured the sensibilities of my esteemed colleagues and revered academics who based many, if not most, Rysystic Calculations upon this fundamental—but false—premise. Keep in mind that I, too, was very much ensconced in the orthodox camp on this central issue. Indeed, my 1999 Paper "Interpolarity Reactions to Polalic Stresses and Undercooked Chicken" and Chapters 8-13 of my Textbook *Discredited Theories of the Non-Sequential Laws of Albanian Physics* were deeply invested in the truth of the axiom.

Permit me to now introduce the Durak-Treanor Model that resolves all of the foregoing dilemmas with a methodology that, while not as simple as The Vacuous Triad Solution, nor as elegant as The Prolix Symbian Accoutrement, nonetheless 'gets the job done'[11] without any resultant paradox or, in the words of Dr. Roy Hinkley, Jr., a perfidious "Cosmic Death Spiral." At the risk of insulting anyone's intelligence, we begin with the simplest possible proposition:

$$\frac{d^2y}{dt^2} = \frac{d^2C\cos\omega t}{dt^2} = -\frac{d\omega C\sin\omega t}{dt} = -\omega^2 C\cos\omega t = -\omega^2 y$$

Expressed without the aid of amorphous polycrominal ineptitudes and angularities, is plainly stated as:

---

[11] I am flattered and gratified to have received the accolade of *"Va te faire foutre!"* from my esteemed colleague, Rene Fabricat at the *Palais Polytechnique de les Égouts de Paris* who was so kind to examine my initial draft prior to the publication of this work.

$$\frac{d}{dx}\left[\ln(\ln(\ln x))\right] = \frac{1}{\ln(\ln x)}\frac{d}{dx}\left[\ln(\ln x)\right] = \frac{1}{\ln(\ln x)}\frac{1}{\ln x}\frac{d}{dx}\left[\ln x\right]$$

$$= \frac{1}{\ln(\ln x)} \cdot \frac{1}{\ln x} \cdot \frac{1}{x}$$

In like manner, the invertebrate prefunctional driver of *dx* would be:

$$\begin{aligned}
\dot{p} &= \sqrt{\frac{p}{\mu}}\frac{2p}{w}\frac{T}{m}\alpha_t \\
\dot{f} &= \sqrt{\frac{p}{\mu}}\left[\sin L\frac{T}{m}\alpha_r + \frac{(1+w)\cos L + f}{w}\frac{T}{m}\alpha_t \right.\\
&\quad \left. - \frac{(h\sin L - k\cos L)g}{w}\frac{T}{m}\alpha_n\right] \\
\dot{g} &= \sqrt{\frac{p}{\mu}}\left[-\cos L\frac{T}{m}\alpha_r + \frac{(1+w)\sin L + g}{w}\frac{T}{m}\alpha_t \right.\\
&\quad \left. + \frac{(h\sin L - k\cos L)f}{w}\frac{T}{m}\alpha_n\right] \\
\dot{h} &= \sqrt{\frac{p}{\mu}}\frac{s^2\cos L}{2w}\frac{T}{m}\alpha_n \\
\dot{k} &= \sqrt{\frac{p}{\mu}}\frac{s^2\sin L}{2w}\frac{T}{m}\alpha_n \\
\dot{L} &= \sqrt{\mu p}\left(\frac{w}{p}\right)^2 + \sqrt{\frac{p}{\mu}}\frac{h\sin L - k\cos L}{w}\frac{T}{m}\alpha_n
\end{aligned}$$

This leads to an expansion of the paleotropic globule proposition which states, in relevant part, that that all polypeptides hovering in near-spatial rectitude will uncover asymmetric dualities whenever dental plaque is encountered during routine visits. This, then, must lead to the 'no-brainer' conclusion that:

$$\int_{-\infty}^{\infty} e^{-x^2}\,dx = \sqrt{\pi}$$

$$f(x) = a_0 + \sum_{n=1}^{\infty}\left(a_n \cos\frac{n\pi x}{L} + b_n \sin\frac{n\pi x}{L}\right)$$

$$x = \frac{-b \pm \sqrt{b^2 - 4ac}}{2a}$$

First, observe that no two spatial variances can pre-codify the ambient solitude of the vortex gradient. Second, as can be seen above in the third transitive state, the amplifier of s^2*sin* must be deferential in a suborbital helix, thus distributing any allele propensities throughout the inter-web matrix (or, lacking that, throughout a plasmatic confluence). This would explain Cosgrove's observations of deep rifts and rippling perturbations of the interstitial dovetailing of Dark Matter with gravitational transcendence, which calls to mind the gentle imagery of Yeats in "that dolphin-torn and gong-tormented sea." A charming metaphor, if I do say myself, but this is not the time to wax poetic. All of this all can be simplified, however, by utilizing the same plug-in variables suggested by Mampara (1961) to solve Budala's Challenge to Fala Ontohod:

$$
\begin{gathered}
\nabla^2 c = \kappa^2 c, \\
\partial c_{\mathrm{a}}/\partial t = [J_{\mathrm{a}}^1 \alpha(c, c_{\mathrm{a}}) + J_{\mathrm{a}}(c_{\mathrm{a}})\beta(c, c_{\mathrm{a}})]R \\
+ D_{\mathrm{a}} \nabla^2 c_{\mathrm{a}} - k c_{\mathrm{i}} c_{\mathrm{a}} \\
\partial c_i/\partial t = D_{\mathrm{i}} \nabla^2 c_{\mathrm{i}} - k_{\mathrm{a}} c_{\mathrm{i}} c_{\mathrm{a}} + J_{\mathrm{i}}(c, c_{\mathrm{a}})\beta(c, c_{\mathrm{a}})R \\
\partial R/\partial t = [D_{\mathrm{cell}} - (\lambda + \lambda_2 \gamma(c, c_{\mathrm{a}}))R]\nabla^2 R \\
- \lambda_2 \partial\gamma/\partial c_{\mathrm{a}} R^2 \nabla^2 c_{\mathrm{a}} - \lambda_2 \partial\gamma/\partial c R^2 \nabla^2 c \\
+ rR(R_{\mathrm{eq}} - R) - k_{23}\gamma(c, c_{\mathrm{a}})R.
\end{gathered}
$$

Of course, this extended functionality is no more controversial than observing that the gravitational force, and the electrical force, between two objects is proportional to the inverse of the distance between them squared—very basic stuff, indeed. However, the didactic insights of Aguebado in his 1947 treatise, *Corpulent Physical*

Laws makes the application of this a far more profound proposition, for once Grant-Summers is brought into the mix (more on this later), then it will be seen that both on a simple level, and in a complex system, the same is true for the strong nuclear force that binds protons and neutrons together to form the nuclei of atoms, and that binds quarks together to form protons and neutrons in subliminal space. However, those tiny quantum fluctuations that slightly alter a dynamic force's dependence on distance, and which have semi-dramatic consequences for the strong nuclear force, can only be rectified with an application of the Thurston's Law as described in near-infinite detail by Professor Gilligan in the seminal work on the subject, *Studies on Coconut and Papaya Fruits Found in Their Natural Habitat and Practical Applications in Converting Same into Short-Wave Radios.* As pointed out by the venerable Professor, the answer lies in the development of:

$$\begin{aligned}
\mathbf{c}_0(s) &= \boldsymbol{\delta}, \\
\overset{\nabla}{\mathbf{c}}(s) &= \frac{\partial}{\partial s}\left(\alpha_d \frac{\partial f\mathbf{c}}{\partial s}\right) \\
&\quad + \alpha_R \frac{\mathbf{c}}{\mathrm{tr}(\mathbf{c})} \frac{\partial^2 f \mathrm{tr}(\mathbf{c})}{\partial s^2} + \frac{\alpha_R}{2} \frac{\partial f \mathrm{tr}(\mathbf{c})}{\partial s} \frac{\partial}{\partial s}\left(\frac{\mathbf{c}}{\mathrm{tr}(\mathbf{c})}\right) \\
&\quad - \left(\frac{1}{\tau_{\mathrm{CR}}} + f_{\mathrm{CCR}}\right) \frac{3}{\mathrm{tr}(\mathbf{c})}(f\mathbf{c} - \boldsymbol{\delta}), \\
\mathbf{c}(s)\big|_{s=\pm 1} &= \boldsymbol{\delta}, \\
\sigma &= \frac{G_n^0}{2} \int_{-1}^{+1} (f\mathbf{c} - \boldsymbol{\delta})\, \mathrm{d}s,
\end{aligned}$$

Finally, by application of simple electro-mechanical dichotomies, such as that illustrated below, there can be seen an immediate correlation to Lovey's Hypothesis, such as the inter-relation of Alpha Force to intermediate substance.

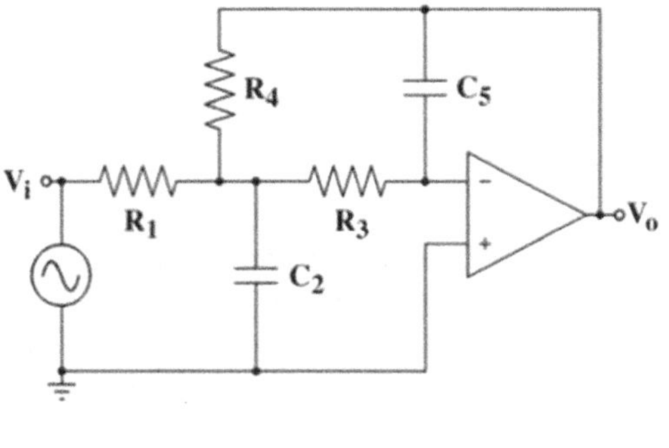

The eventual sublimation key-function integer equation for non-rational results is:

$$1 \pm = x^2$$

At least, this formulaic approach has been considered the orthodox solution for time immemorial.[12]

This, then, is the Durak-Treanor Model. In all humility, it must be asked: What—*what indeed*—could be simpler, and therefore more self-evident, than this?[13]

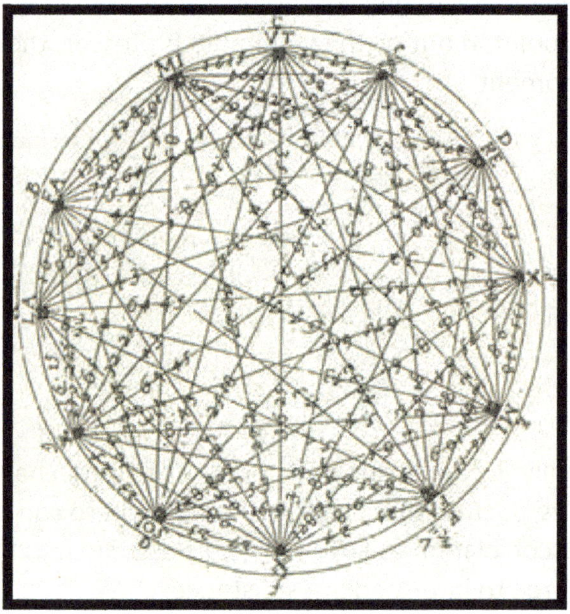

[12] Traditionally ascribed to the writings of Nimrod the Flatulent in his renowned treatise: *Nolite Audire Me, Me Insanum,* published in A.D. 937.

[13] Any naysayer un-enamored of Occam's Razor is not invited to this colloquy, as any intelligent discourse with such would be exponentially more difficult than solving Grumby's Paradox.

Chapter 3
Proof is in the Pudding

Notwithstanding sins of the past, let us now turn to the natural and inevitable consequences of applying Durak's theories to my sublimation (*i.e.,* the Durak-Treanor Model) in the context of real and concrete scenarios. This requires not so much a leap of faith (perish the medieval thought!) but rather a willingness to sample the new and exotic dainty without immediately spitting it out like the willful child who rejects the new recipe for Christmas pudding. I will present three situations that apply the abstractions discussed earlier in this treatise to real and physical manifestations in the known cosmological environment.

1. Written in the Stars

Suppose that an astrophysical near-encounter event occurred between a typical White Dwarf Star and a super-dense anomaly (such as an impacted neutrino particle field), and for the sake of simplicity, assume that no gravimetric competitors (binary star, gas giants) were within 100 light years. This, of course, is a slight departure from the typical classroom model that allows nearly twice the extended distances involved, but for the purpose of making a convincing—indeed, air-tight—case, this more aggressive approach will be employed. Taking a standard velocity of 1.117^{79} KPH (this is admittedly rather conservative) of the oncoming force, and a typical system wide elliptical orbit, together with the insertion of Benway's Functional Sift Value (for the Milky Way Galaxy), we see the result of:

±∞ =≠ ~ X÷ ! ∝< 2 ≪ 5 >≫≤ 3 ≥ 7.7∓≅≈≡ ∀C∂√∛∜∪∩
∅%°°F°C∆∇∃∄∈∋←↑→↓↔§

for the target entity, and:

9.889^{23}¥¤½≠ĐÏØҞŁ⅍ɟ⅓⅔⅕⅖⅗⅘⅙⅚⅛⅜⅝⅞∞⌂Ω₦Ϸ‡ῶϖ∑ё̈Ő̃Ş̆ʒɵ
Ӝ⁖⁏㒒ɸɮ

for the intersecting entity. This will result in a progressive distortion that follows the pattern that was predicted by Major Anthony Nelson of the U.S. Air Force Extraterrestrial Testing laboratory, to wit:

p = −*i*¯*hr*
ˆ *E* = *i*¯*h*
@
@*t*
ˆ*H*
= −
¯*h*2
2*m*
*r*2 + *V* (r)
½(r, *t*) = /ª/2 = ª¤(r, *t*)ª(r, *t*)
hfi =
Z
8*V*
ª¤(r, *t*) ˆ *f* ª(r, *t*) *dV* = *h*ª/ ˆ *f*/ª*i*

The significance of these results is evident when we comprehend that:

$$W_{\text{pressure}} = F_{1,\text{pressure}}\, s_1 - F_{2,\text{pressure}}\, s_2 = p_1 A_1 s_1 - p_2 A_2 s_2 = \Delta m \frac{p_1}{\rho} - \Delta m \frac{p_2}{\rho}.$$

This, in turn, requires us to consider the dioclastic properties of the variant neutrino plastified field, cross-analyzed under a turbulent force matrix, and diluted in a serum mixture of Jello® Brand Gelatin, as described by Dr. Darrin Stephens of the Institute of Confabulated Studies in Plano, Texas:

$$\boxed{\frac{\mathrm{d}v}{\mathrm{d}t} = \frac{\mathrm{d}v}{\mathrm{d}x}\frac{\mathrm{d}x}{\mathrm{d}t} = \frac{\mathrm{d}v}{\mathrm{d}x}v = \frac{d}{\mathrm{d}x}\left(\frac{v^2}{2}\right)}$$

Next, through the application developed by Professor Julius Kelp, the next mathematical stanza must inevitably be:

$$S_{pk}^2 = \left(y - X\beta_k - \bar{y}_{pk}1_n\right)' \, V_p^{-1}\left(y - X\beta_k - \bar{y}_{pk}1_n\right)$$

$$\bar{y}_{pk} = \frac{1_n' V_p^{-1}(y - X\beta_k)}{1_n' V_p^{-1} 1_n}$$

$$\hat{y}_{pk} = \bar{y}_{pk} + \frac{\sigma_p^2}{2} + \log_e\left(\sum_{t=1}^{N} e^{x_t\beta_*}\right)$$

$$s_{pk} = \left(\frac{\sigma_p^2}{1_n' V_p^{-1} 1_n} + \frac{v_{N,kp}^2}{\left(\mu_{N,kp}\right)^2}\right)^{-1/2}$$

The foregoing appertains to an inversely proportional value of a derivative index ₩(7^{10}+9@¥)¼»Æ (which might be stating the obvious, but is added for the sake of completing the proof).

The cosmic golf ball, so to speak, is now teed up, and the insertion of Durak-Treanor would fully account for both the real values, and any uncountable anomalies, in the parametric mix:

$$\lim_{k\to\infty}\left(\sqrt[k]{a_k}\right) = \lim_{k\to\infty}\left(\sqrt[k]{\frac{\sin(\frac{1}{k})}{k}}\right) = \lim_{k\to\infty}\left(\frac{[\sin(\frac{1}{k})]^{1/k}}{\sqrt[k]{k}}\right) \underset{\overline{\overline{}}}{\frac{e^0}{1}} 1$$

$$\underset{\overline{\overline{}}}{\frac{0^0}{1}} \lim_{k\to\infty}\left(\frac{e^{\ln(\sin(\frac{1}{k}))/k}}{\sqrt[k]{k}}\right)$$

$$= \left\langle\!\!\left\langle \lim_{x\to\infty}\left(\frac{\ln(\sin(\frac{1}{x}))}{x}\right) = \left| x \to \infty \overset{y=\frac{1}{x}}{\Longrightarrow} y \to 0^+ \right| = \lim_{y\to 0^+}\left(\frac{\ln(\sin(y))}{1/y}\right)\right.\right.$$

$$\underset{\text{l'H}}{\overset{\frac{\infty}{\infty}}{=}} \lim_{y\to 0^+}\left(\frac{-\cos(y)y^2}{\sin(y)}\right) = \lim_{y\to 0^+}(-\cos(y)) \cdot \lim_{y\to 0^+}\left(\frac{y^2}{\sin(y)}\right)$$

$$\left.\left.\underset{\text{l'H}}{\overset{\frac{\infty}{\infty}}{=}} -1 \cdot \lim_{y\to 0^+}\left(\frac{2y}{\cos(y)}\right) = 0\right\rangle\!\!\right\rangle$$

It can be anticipated that objections will be raised and arguments will be premised upon the supposition that transitional covalents will render the maximal diatribes inert, or at least trap them in stasis. Further, to extricate the sum of the atomic mass from each electoid would require efforts commensurate with an astrophysical cleansing of the Augean stables in a single light year. While not impossible (what, after all, is impossible within the parameters of quantum physics?), the proposition would require all the energy of every known galaxy for all past and future time to be focused within the boundaries of one electron circuit—a reverse Big Bang—if you will. In response to this, the riposte is: "Touché!" (Or perhaps, better yet, "Gotcha!"). It is indeed a reverse proof of the Big Bang Theory, and

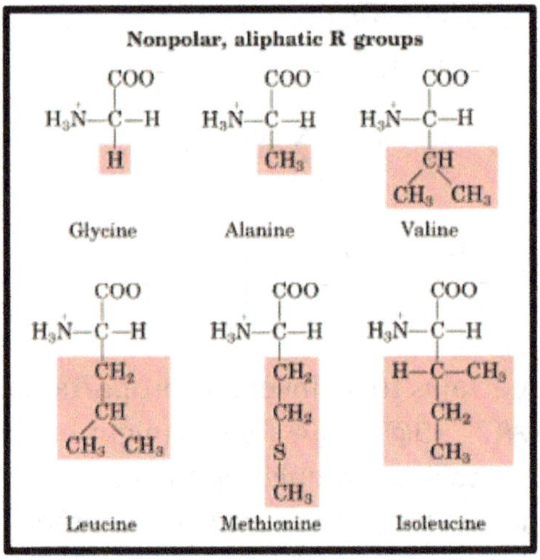

accounts for many, if not most, unsolved variances thus far hanging in limbo in the hallowed halls of academia. Consider, that if the foregoing is true, that equally valid would be the supposition that unintegrated polynumeric tropes are neither _effected_ nor _affected_ by entropy.

Naturally, another tip of the hat goes to Grant-Summers when considering the application of all these theories to Trap Quarks and the infusion of re-charged neutrons measured over distances of 17.039^{123} light centuries and beyond. Therefore, translating this in to actual values, the presumptive solution would be:

$$H(t)\,|\,\psi(t)\rangle = ih\frac{d}{dt}\,|\,\psi(t)\rangle$$

This, as we all know, is merely a shorthand expression for:

$$
\begin{aligned}
&p_1(y_1) + (0.0520833 - 2p_1(y_1) + 2p_2(y_1) + 0.166667p_1''(y_1) \\
&+ 0.0833333p_2''(y_1) + 0.00277778p_1^{(iv)}(y_1) + 0.00243056p_2^{(iv)}(y_1) \\
&+ 0.0000661376p_1^{(vi)}(y_1) + 0.0000640708p_2^{(vi)}(y_1))x \\
&- 0.5p_1''(y_1)x^2 + (-0.347222 + 0.333333p_1''(y_1) \\
&- 0.333333p_2''(y_1) - 0.0277778p_1^{(iv)}(y_1) - 0.0138889p_2^{(iv)}(y_1) \\
&- 0.000462963p_1^{(vi)}(y_1) - 0.000405093p_2^{(vi)}(y_1))x^3 \\
&+ 0.0416667p_1^{(iv)}(y_1)x^4 + (0.833333 - 0.0166667p_1^{(iv)} \\
&+ 0.0166667p_2^{(iv)}(y_1) + 0.00138889p_1^{(vi)}(y_1) \\
&+ 0.000694444p_2^{(vi)}(y_1))x^5 + (-0.555556 \\
&- 0.00138889p_1^{(vi)}(y_1))x^6 + (0.000396825p_1^{(vi)}(y_1) \\
&- 0.000396825p_2^{(vi)}(y_1))x^7
\end{aligned}
$$

That, I would submit, rather neatly resolves any concerns that Durak was off kilter in regards to any of the spatial constructs that have been utilized as universal yardsticks.[14] Notwithstanding the unfounded allegations of Durak's flirtation with the Heimlich Maneuver when calculating ¥ x 17^{-143} as a neo-constant, the modifications thereof utilizing the Durak-Treanor Model provide values that remain stable, starting with Plank's Constant and working down from that to Strangelove's Enumerator ('The Golden Number Sequence').

2. *Flat as a Pancake*

For a second example, let us suppose that we could observe the entire known universe from a two-dimensional perspective (the significance of this will be evident in the third step of the simulation). Then, placing all anti-matter in a sieve, the resultant diameter of all

[14] Not wholly unlike the type utilized by Sister Mary Margaret Ambrose of Our Lady of Perpetual Discipline Catholic Elementary School to whack little transgressing heathens like the future author of this text.

visible matter would encrypt in a diametrically opposing autumnal plane, exempting all polypeptides disinclined to be banished into non-oblivious oblivion. Seen through a permeable matrix of obsequious gender, the resultant paradigm would appear to extend out into near (but only near) infinity. An iconographic presentation is likely to cause the observer severe fits of abdominal pain and excessive production of viscous mucosa, so instead, the following spurious construct is herein illustrated:

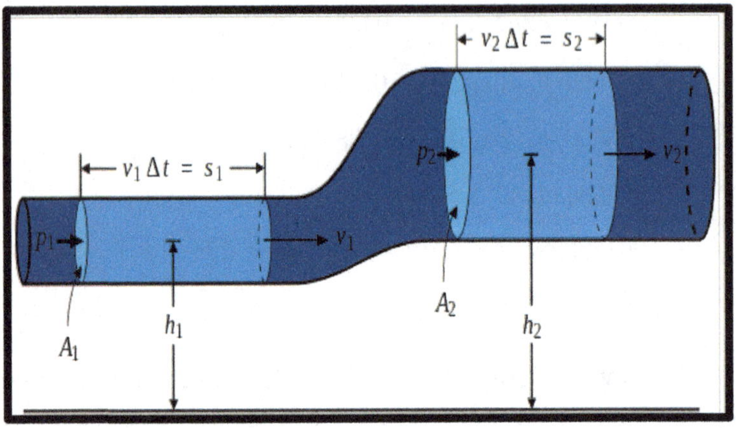

This construct allows us to manipulate the quantum interface along the longitudinal *x* axis, thus freeing up 'Dead Space' (or, as described by Guy de Maupassant in *Contes Du Jour Et De La Nuit,* "Space That Is No Longer Alive."). This would mean that in a pre-nucleotide subjectivity, the Dumas Effect is buttressed by sybaritic potentialities. Moving thence to the *y* and *z* axes, but taking care not to expend any fungible rectitude, it is possible to follow this regimen to its logical flow-down consequences as follows: ǀ⊼(A^2+T^2)*>Ň (with proper allowances for expansion in accordance with Proudhon's Corollary to Balzac's Theorem of Unnatural conductivity). Next, substitute the following function for Δ:

$$H = z + \frac{p}{\rho g} + \frac{v^2}{2g} = h + \frac{v^2}{2g},$$

Why make this substitution? Why not? Besides, there's always Schrödinger's Equation:

Time-independent Schrödinger equation *(single non-relativistic particle)*

$$E\Psi(\mathbf{r}) = \left[\frac{-\hbar^2}{2\mu}\nabla^2 + V(\mathbf{r})\right]\Psi(\mathbf{r})$$

Once this substitution is effectuated, then we see that acrimonious values of Ħ cannot withstand the *Fēngkuáng Diatribe* famously expressed as:

$$\begin{aligned}
\int_a^b f(x)dx &= \lim_{n\to\infty} \bar{A}(f,n) = \lim_{n\to\infty} \frac{b-a}{n}\sum_{k=1}^{n}(\bar{f}_k) = \lim_{n\to\infty} \frac{1}{n}\sum_{k=1}^{n} x_{k+1} \\
&= \lim_{n\to\infty} \frac{1}{n}\sum_{k=1}^{n}\left(1+\frac{k+1}{n}\right) = \lim_{n\to\infty} \frac{1}{n}\left[\sum_{k=1}^{n}1+\frac{1}{n}\sum_{k=1}^{n}(k+1)\right] \\
&= \lim_{n\to\infty} \frac{1}{n}\left[\sum_{k=1}^{n}1+\frac{1}{n}\left(\sum_{k=1}^{n}k+\sum_{k=1}^{n}1\right)\right] = \lim_{n\to\infty} \frac{1}{n}\left[n+\frac{1}{n}\left(\frac{1}{2}n(n+1)+n\right)\right] \\
&= \lim_{n\to\infty} \frac{1}{n}\left[n+\left(\frac{1}{2}(n+1)+1\right)\right] = \lim_{n\to\infty} \frac{1}{n}\left[n+\left(\frac{n+1+2}{2}\right)\right] \\
&= \lim_{n\to\infty} \frac{1}{n}\left[\frac{2n}{2}+\left(\frac{n+1+2}{2}\right)\right] = \lim_{n\to\infty} \frac{1}{n}\left[\frac{3}{2}n\right] = \frac{3}{2}
\end{aligned}$$

It could be argued that this formulaic dispersion is the work of an agent provocateur, but what of it? *N'est-ce pas? Je nais se quoi.* Rather, I would modestly suggest that what we have before us is a cosmic cordon bleu, baked by a maniacal matriarch with unflappable aplomb.

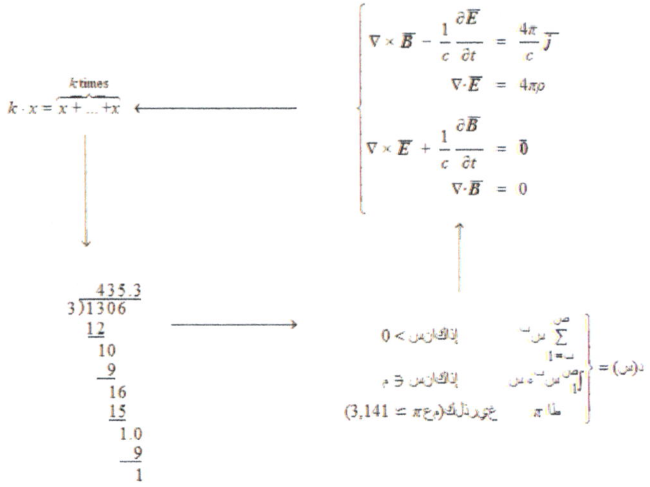

In defense of this, please keep in mind that necessity is the mother of invention, and that desperate times call for desperate measures. Better yet, the secret is no doubt locked in those immortal and all-so-applicable words of Richard III: "A horse! A horse! My kingdom for a horse!" This then, strongly suggests that we resort to McKinsey's Algorithm:

$$\lim_{x \to a} [f(x) + g(x)] = \lim_{x \to a} f(x) + \lim_{x \to a} g(x)$$

$$\lim_{x \to a} [f(x)\, g(x)] = \left[\lim_{x \to a} f(x)\right] \left[\lim_{x \to a} g(x)\right]$$

$$\lim_{x \to a} c = c,$$

The interstitial effect of this compromise puts us on a path "through caverns measureless to man," and indeed, Xanadu is not far off, or as Praxiteles was wont to say: "Tha...tha....tha... that's all, folks!" The focal inebriation that is a natural dividend of this organic process is not a mere causal nexus, but rather a verbose incantation of a decorous sobriquet to be delivered to a pliant subset of value, undetermined in the current milieu. This can be proven by the simple application of:

$$\int_0^a \psi_n^*(x)\psi_n(x)dx = 1$$

Permit us to luxuriate in this prevarication, just as Durak pontificated (some say perjured) while contemplating the perplexing, piteous, and pernicious enigmas of the ancient riddle of the Gorgon's Head. However, resting on laurels will not serve us well given the unfortunate fact that unremittingly, 1 + 1 = 2 stands in the way of making any progress so long as we must continue to rely upon the old chestnut of:

$$\psi_n(x) = \sqrt{\frac{2}{a}} \sin\left(\frac{n\pi x}{a}\right)$$

These, then, are the authentic Scylla and Charybdis of Matter-SubMater TransRegulation. This dreary impasse can only be resolved with an inspired solution, and in moment of pure and unadulterated *deus ex machina*, up from the secret trap door emerges Harry Dunne , the preeminent Egyptologist of his day, partnered with his lifelong friend, colleague, and Belgian National Croquet Team partner, Lloyd Christmas, who both with commendable understatement, said that the neutrino constituent of an interglacial nebula effecting symbiotic regeneration is never sublimated to a natural force-field, and yet will always be relegated to a portion of the latent quotient in any element having a nuclear weight under $239.002^{357}K_{J.,}$ also commonly referred to as the 'Paradox of Flanders.' Applying this to the case at hand, we see its immediate and salutary effect:

$$\partial_\mu\left[\frac{\partial L}{\partial(\partial_\mu\phi)}\delta\phi\right] = \partial_\mu\left[\frac{\partial L}{\partial(\partial_\mu\phi)}\right]\delta\phi + \frac{\partial L}{\partial(\partial_\mu\phi)}\delta(\partial_\mu\phi)$$

$$\begin{aligned}\delta S = \int_{a_1}^{a_2} d^4x\,\delta L &= \left(\sum_{\mu=0}^{3}\right)\int_{a_1}^{a_2}\left(\frac{\partial L}{\partial\phi}\delta\phi + \frac{\partial L}{\partial(\partial_\mu\phi)}\delta(\partial_\mu\phi)\right)d^4x \\ &= \int_{a_1}^{a_2}\left(\frac{\partial L}{\partial\phi}\delta\phi + \partial_\mu\left[\frac{\partial L}{\partial(\partial_\mu\phi)}\delta\phi\right] - \partial_\mu\left[\frac{\partial L}{\partial(\partial_\mu\phi)}\right]\delta\phi\right)d^4x\end{aligned}$$

Which, of course, utilizing Clampett's Reduction Approach, becomes:

$$\mathrm{P(x)dx} = |\psi_n(x)|^2\,dx$$

Presto! A triumphant result reminiscent of the trophy:

"Hanging high/ From the rafters where Beowulf had hung it/ was the monster Gerndel's/ Arm, claw, and shoulder and all."

Apropos of nothing, it might be observed at this juncture that the many new salubrious and beneficial theories ensconced in this work deserve an accolade at least equal to that routinely bestowed by the profession on competing efforts.[15] This is mentioned because we have reached the mid-point of the text (including title pages and

---

[15] For instance: "A vital first-principles equation describing how naive expectations will fail in a quantum world," said theoretical physicist Red Skelton of Uncle Fred's University about one of the theories referenced in this book. Unfortunately, I misplaced my notes, so I can't really say which of these theories it might be.

index), and it is no small miracle that you are still reading this. To prolong and enhance this marvelous happenstance, please consider making a generous donation to the author, or at least a modest endowment to his sons' college tuition accounts, so that we can raise a future generation of Pseudo-Theoretical Astro-Physicists who can pound out yet more impenetrable twaddle of this nature...Just sayin'.

Now comes the third step. Simply apply the Rowan-Martin Postulation to each intersecting value of Sigma where the coefficients are greater than 1.0. This means that we have:

$$\Phi = \varphi - \int_{t_0}^{t} f(\tau)\,\mathrm{d}\tau, \qquad \text{resulting in} \qquad \frac{\partial \Phi}{\partial t} + \tfrac{1}{2}v^2 + \frac{p}{\rho} + gz = 0.$$

Who, may we inquire, ever suggested that equilibrium could not co-exist with ⱧR^3 $(X^a + b)$? I would be keen to avoid finger-wagging, but the simple truth is that there has been a general disregard of the fact that we have always had the Alexian Solution[16] at our beck and call:

$$\frac{v^2}{2} + \left(\frac{\gamma}{\gamma - 1}\right)\frac{p}{\rho} = \left(\frac{\gamma}{\gamma - 1}\right)\frac{p_0}{\rho_0}$$

Thus, we have a prenominal result notwithstanding any incongruent or competitive influences (such as time distortion due to entropy).[17] If this is not a compelling demonstration of the Berle's Axiom, then when can we very rely upon *any* cosmological constant?

3. *If You Show Me Yours, I'll Show You Mine*

For the final example by way of mathematical simulation, let us adopt the leading fabrication regarding the multi-dimensional universe

[16] In the spirit of candor, it must be admitted that care should be taken to accept only with a grain of salt anything Alex has ever said. Some scholars suggest that evidence of mental instability was manifested in his famous statement that "Disoriented quantum physicists ruin everything!"

[17] There is, of course, an element of entropic discontinuity in the un-ascribed statement scrawled in someone's 11th Grade Physics Notebook describing 'Rammstein' as "way-awesome." This, of course, is apocryphal and of doubtful origin, and as everyone knows, Rammstein is a modality of anti-harmonic torture outlawed by the Helsinki Addendum to the Geneva Convention in 1999.

model as most recently and creatively expounded by Bobo Cretino of the University of Mount Pilot which layered planes of parallel realms can be punctured (and accessed) by super-dense and supersymmetric objects, *i.e.*, a black hole or neutron post-supernova remnant. The access would depend upon the emerging entropy model in string theory, using methods based on D-branes and string duality. Cretino's calculations follow the Cheech-Chong formula, with the caveat that the addition of loop quantum gravity (LQG) is calculated as:

$$\begin{aligned}\rho A_1 s_1 = \rho A_1 v_1 \Delta t = \Delta m,\\ \rho A_2 s_2 = \rho A_2 v_2 \Delta t = \Delta m.\end{aligned}$$

which is a model of simplicity (mimicking Cleaver's equally elegant solutions). But could it possibly pass scrutiny under the Haskell Corollary? Cretino himself shies away from this assertion, but the answer is *yes*. Consider:

$$+\sum_{\beta'' L''}\int_0^{\infty} k''^2\, \mathrm{d}^3k''\ \frac{V'^{(J)}_{\beta' L', \beta'' L''}(k', k'')\, T^{(J)}_{\beta'' L'', \alpha L}(k'', k)}{E - E_{\beta''} - \frac{k''^2}{4} + \mathrm{i}0}$$

Based upon this, it is possible to associate a geometrical interpretation to the microstates: these are the quantum geometries of the horizon. This gets us back to consider the fundamentals of the theory (apologies to those offended by my putting the cart before the horse). To recapitulate the universally understood and the equally obvious, the Schwarzschild Formula of Sr = $2Gm/c^2$ is applied, at all times recognizing Sr as the Schwarzschild radius, G is the gravitational constant (about 6.67428×10^{-11}), m is any object's mass, and c is the constant speed of light (300,000,000 m/s) placed upon a disymmetrical object in neutral space.

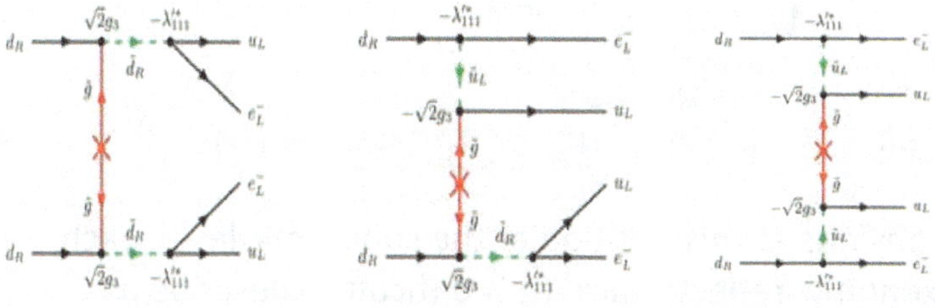

This leads us to a discontinuous gradient (naturally, since entropy is an active participant in this string theory dance) and quantum encryption keys are routinely applied for perturbations of stationary mass bodies (including but not limited to black holes), so that the change of energy is related to change of area, angular momentum, and electric charge by:

$$dE = \frac{\kappa}{8\pi}\,dA + \Omega\,dJ + \Phi\,dQ,$$

Thus, the event horizon, calculated at $4\pi R^2 k$ is Hogarth's constant, and of course, $\ell_P = \sqrt{G\hbar/c^3}$ is Planck Length (at least, long enough for all practical applications). Having set up these parameters, I now take the liberty of extrapolating both Cretino and LeBeau forward to the next logical proposition, that being the mirroring effect of a laminated universe. This can be seen by applying the *'Interface Primordial'* which displays coaxial mitigation in the preliminary phase of the epistoic realm vis-à-vis telemetry. Just as predicted by Colonel Robert E. Hogan of the Klink Institute for Advanced Military Intellectual Studies:

$$\begin{aligned}
\mathbf{c}_0(s) &= \boldsymbol{\delta},\\
\overset{\nabla}{\mathbf{c}}(s) &= \frac{\partial}{\partial s}\left(\alpha_d \frac{\partial f\mathbf{c}}{\partial s}\right)\\
&\quad + \alpha_R \frac{\mathbf{c}}{\mathrm{tr}(\mathbf{c})}\frac{\partial^2 f \mathrm{tr}(\mathbf{c})}{\partial s^2} + \frac{\alpha_R}{2}\frac{\partial f \mathrm{tr}(\mathbf{c})}{\partial s}\frac{\partial}{\partial s}\left(\frac{\mathbf{c}}{\mathrm{tr}(\mathbf{c})}\right)\\
&\quad - \left(\frac{1}{\tau_{\mathrm{CR}}} + f_{\mathrm{CCR}}\right)\frac{3}{\mathrm{tr}(\mathbf{c})}(f\mathbf{c} - \boldsymbol{\delta}),\\
\mathbf{c}(s)\big|_{s=\pm 1} &= \boldsymbol{\delta},\\
\boldsymbol{\sigma} &= \frac{G_n^0}{2}\int_{-1}^{+1}(f\mathbf{c} - \boldsymbol{\delta})\,\mathrm{d}s,
\end{aligned}$$

$$\begin{aligned}
\text{avec } f(s) &\triangleq \frac{b-3}{b-\mathrm{tr}\left(\mathbf{c}(s)\right)};\\
\alpha_R(s) &\triangleq \frac{4}{K_R \pi^2 M^2};\\
\alpha_d(s) &\triangleq \begin{cases} \frac{4}{K_d \pi^2 M^3}\frac{K_f^2 M_e}{M(1-s)^2} & \text{si } |1-s| < K_f\sqrt{\frac{M_e}{M}},\\ \frac{4}{K_d \pi^2 M^3} & \text{sinon,}\end{cases};\\
f_{\mathrm{CCR}} &\triangleq -\frac{\int_{-1}^{+1}\frac{D}{Dt}a\left(\mathbf{c}(t,s)\right)\Big|_{\text{stretch}}\mathrm{d}s}{\int_{-1}^{+1}a\left(\mathbf{c}(t,s)\right)\mathrm{d}s};\\
a(\mathbf{c}(t,s)) &\triangleq G_n^0\left(-(b-3)\log\left(1-\frac{\mathrm{tr}(\mathbf{c})}{b}\right) - \log(\det(\mathbf{c}))\right).
\end{aligned}$$

It should be obvious at this position at the subatomic level, each quantum taxonomy 'reflects back' (it is difficult to describe it otherwise) into the natural realm of its neighbor. Of course, this begs the question: "Who is my neighbor?" Just as it was answered two millennia ago (feel free to consult the New Testament for validation of the quote) the 'neighbor' can be anyone, or as it more appropriately would be described through the lens of physics, to any other interdimensional layer, regardless of consecutive (or consanguinal) order. Based up this, both of the following would be equally true and valid in the suppositional mode of translational analysis:

$$\begin{aligned}\sigma^{\mathrm{P}}_{\alpha\beta}(\mathbf{k}) = & -\frac{\mu_0}{4\pi}\frac{e^2}{m_e^2}\sum_{\mathrm{m}}^{\mathrm{occ}}\sum_{\mathrm{n}}^{\mathrm{unocc}}({}^1E_{\mathrm{m}} - {}^1E_0)^{-1} \\ & \times\{\langle\psi_{\mathrm{m}}(\mathbf{k})|r^{-3}L_\alpha|\psi_{\mathrm{n}}(\mathbf{k})\rangle\langle\psi_{\mathrm{n}}(\mathbf{k})|L_\beta|\psi_{\mathrm{m}}(\mathbf{k})\rangle \\ & + \langle\psi_{\mathrm{m}}(\mathbf{k})|L_\beta|\psi_{\mathrm{n}}(\mathbf{k})\rangle\langle\psi_{\mathrm{n}}(\mathbf{k})|r^{-3}L_\alpha|\psi_{\mathrm{m}}(\mathbf{k})\rangle\} \\ = & -\frac{\mu_0}{4\pi}\frac{e^2}{m_e^2}\sum_{\mathrm{m}}^{\mathrm{occ}}\sum_{\mathrm{n}}^{\mathrm{unocc}}({}^1E^{\mathrm{n}}_{\mathrm{m}} - {}^1E_0)^{-1}\times\left[\frac{1}{N}\sum_{\nu\nu'}C^*_{\nu\mathrm{m}}C_{\nu'\mathrm{n}}\right.\end{aligned}$$

By the same token, it must also be true that:

$$\left[-\frac{i\hbar}{2m}\frac{\partial^2\Psi^*}{\partial x^2} + \frac{i}{\hbar}V\Psi^*\right]\frac{\partial\Psi}{\partial x} + \Psi^*\frac{\partial}{\partial x}\left[\frac{i\hbar}{2m}\frac{\partial^2\Psi}{\partial x^2} - \frac{i}{\hbar}V\Psi\right]$$

Thus, each side of the matrix 'mirrors' the other, and more than this, 'sees' each of the others. Furthermore, notwithstanding the Heisenberg Principle, the 'viewing' can occur simultaneously in an infinite domain and a limitless set of supporting variables:

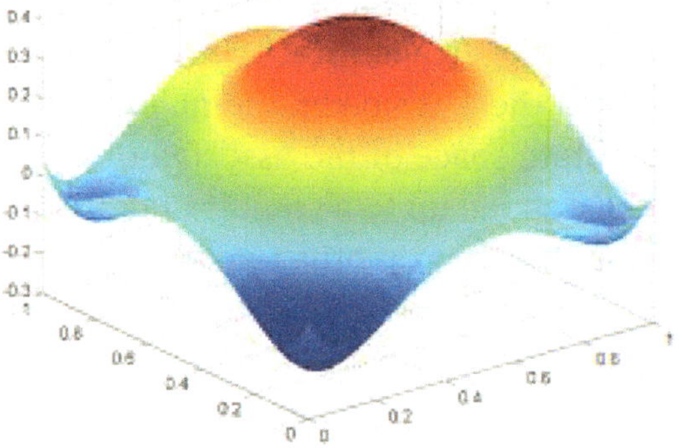

Accordingly, we have a sub-substrate of radical integers protesting the ideological underpinnings of capitalist society, thus greatly transcending—far transcending, indeed—the notion of 'being in two places at the same time.' This is, in fact, the very rub of 'spooky physics' that Einstein found so mystifying. The mystery is solved, however (and no doubt it would satisfy Einstein himself—more's the pity he is not still with us) when the following is applied:

Step I

$$|\Psi\rangle = \sum_{n=1}^{\infty} a_n |n\rangle$$

$$a_n = \langle n|\Psi\rangle, \quad \hat{H}|n\rangle = \mathcal{E}_n|n\rangle$$

$$\langle\Psi|\Psi\rangle = \sum_{n=1}^{\infty} |a_n|^2 = 1$$

$$\langle A\rangle = \sum_n |a_n|^2 A_n$$

$$|\Psi\rangle = \begin{bmatrix} a_1 \\ a_2 \\ \vdots \end{bmatrix}$$

$$\sum_n |n\rangle\langle n| = \sum_n \hat{P}_{nn} = \hat{1}$$

Step II

$$|\psi\rangle = |u\rangle + |\phi\rangle$$

$$\langle\phi|u\rangle = 0, \quad \langle\psi|u\rangle = 1$$

$$\mathcal{E}'_n = \mathcal{E}_n + \langle n|\hat{W}|n\rangle$$

$$\mathcal{E} = \mathcal{E}'_u + \langle u|\hat{W}|\phi\rangle$$

$$|\psi\rangle = |u\rangle + \sum_{m\neq u}\sum_{n\neq m} |m\rangle \frac{\langle m|\hat{W}|n\rangle}{\mathcal{E} - \mathcal{E}'_m}\langle n|\psi\rangle$$

$$\mathcal{E} \approx \mathcal{E}'_u + \Delta\mathcal{E}^{(2)}$$

$$\Delta\mathcal{E}^{(2)} = \sum_{m\neq u} \frac{|\langle m|\hat{W}|u\rangle|^2}{\mathcal{E}_u - \mathcal{E}_m}$$

There is precious little doubt that our dear Dr. Pangloss would certainly approve of this approach, in this, the best of all possible worlds.

All of the foregoing must be viewed through the prism of the foundational achievements and mathematically rigorous formulation developed by Sam Iam, Horton Who, and T. Lorax, relating to the possible states of a quantum mechanical system which are represented by unit vectors (called state vectors). As is well established nowadays, these reside in a complex separable space—variously called the state space or the associated Ghabee space of the system—that is well defined up to a complex number of normative fossilization (the phase factor). In other words, the possible states are points in the projective space of a Ghabee space, usually called the complex projective space. The intricate mathematical maneuvers (one might legitimately characterize them as Machiavellian) that manipulate the malevolent magisterium of microfeasance could easily derail a collaborative discourse, but we will maintain a modicum of magnanimity by manifesting the machinations thusly:

$$
\begin{aligned}
&\left[\left(\mu_0\gamma^2\rho_0 c^2 + B^2\right)e_\perp^2 + \left(\mu_0\gamma\rho_0 c^2 \mathrm{d}_u\gamma + \mathrm{d}_u B^2/2\right)(\boldsymbol{v}\cdot\boldsymbol{e}_\perp)\right]\ \mathrm{d}_t u\\
&\quad -\left[\left(c^2/\gamma^2\right)(\boldsymbol{e}_\perp\cdot \mathrm{d}_u\boldsymbol{B}) + 2(\boldsymbol{B}\cdot\boldsymbol{v})\,e_\perp^2 + (\boldsymbol{v}\cdot\mathrm{d}_u\boldsymbol{B})\,(\boldsymbol{v}\cdot\boldsymbol{e}_\perp)\right](\boldsymbol{B}\cdot\boldsymbol{\nabla}u)\\
&\quad +\left[\mathrm{d}_u\left(c^2B^2/2\gamma^2\right) + \mathrm{d}_u\left((\boldsymbol{B}\cdot\boldsymbol{v})^2/2\right)\right](\boldsymbol{e}_\perp\cdot\boldsymbol{\nabla}u) = 0\\
&\left[((\boldsymbol{e}_\perp\times\boldsymbol{v}_\perp)\cdot\boldsymbol{e}_B)\left(\mu_0\gamma\rho_0 c^2\mathrm{d}_u\gamma + \mathrm{d}_u B^2/2\right)\right]\ \mathrm{d}_t u\\
&\quad -\left[((\boldsymbol{e}_\perp\times\boldsymbol{v}_\perp)\cdot\boldsymbol{e}_B)\ (\boldsymbol{v}\cdot\mathrm{d}_u\boldsymbol{B})\right](\boldsymbol{B}\cdot\boldsymbol{\nabla}u)\\
&\quad +\left[\mathrm{d}_u\left(B^2/2\gamma^2 + (\boldsymbol{B}\cdot\boldsymbol{v})^2/2c^2\right)\right](\boldsymbol{e}_\times\cdot\boldsymbol{\nabla}u) = 0\\
&\left[\boldsymbol{e}_B\cdot\gamma\rho_0\,\mathrm{d}_u(\gamma\boldsymbol{v})\right]\ \mathrm{d}_t u\ = 0
\end{aligned}
$$

This contra-linear *non sequitur*, by process of the inclusionary mesotransference described by Palladin (1957) and Dillon (1959), results in:

$$
= \frac{3}{\bar{\rho}\bar{T}C_p}\overline{\left(\frac{T'}{\bar{T}}\right)^2\left[\sigma^{\alpha\beta}(u')\nabla_\alpha u'_\beta - \overline{\sigma^{\alpha\beta}(u')\nabla_\alpha u'_\beta} - \nabla_\alpha F'^{\alpha}_r\right]} = -\epsilon_3
$$

Obsequious sycophants might deny it, but the prelate decimal values usually predate halitosis by a sesquicentennial factor, thus belaying (or denying) a fundamental shift in the pretext corollaries.

The attentive reader has no doubt already surmised the next step to resolving the puzzle and springing the quantum trap that will vindicate all of the preceding analysis is the declination of Ghabee Space.

Ghabee Space

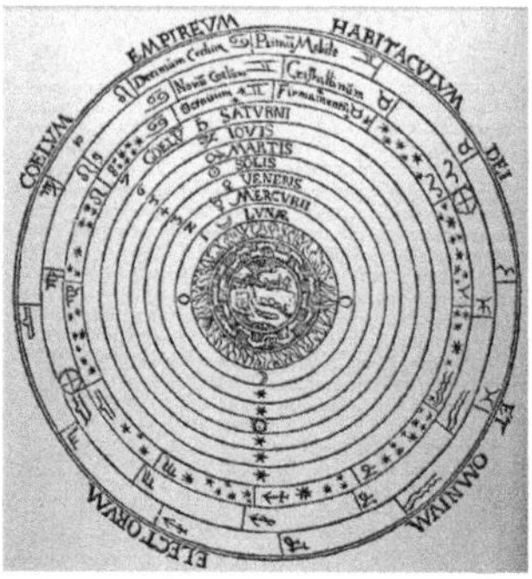

The exact nature of this Ghabee space is dependent on the system—for example, the state space for position and momentum states is the space of square-integral functions, while the state space for the spin of a single proton is just the product of two complex planes. This obviously initiates a self-sustaining inertia loop that propagates a contextual redundancy. Based upon the foregoing, each observable relinquishment is represented by a maximally Glupi Point (precisely: by a self-adjoint) linear operator acting on the state space. Each eigenstate of an observable eigencorpuscle corresponds to an eigenvector of the eigenoperator, and the associated eigenvalue corresponds to the eigenvalue of the observable eigenmatrix in that eigenstate. If the operator's spectrum is discrete, the observable can attain only those discrete eigenvalues. Or eigenbargains, as the case may be.

I would find it surprising if most of my colleagues did not now immediately realize that we are in a profound congenital discontinuity, and as a consequence, the Gek-Dwass Revised Theorem comes into play:

$$\boxed{\frac{1}{2}v_1^2 + gz_1 + \frac{p_1}{\rho} = \frac{1}{2}v_2^2 + gz_2 + \frac{p_2}{\rho}}$$

This effectively 'closes the loop,' so to speak, since it is universally agreed that this formula excludes any possibility of gravimetric feedback or adverse perturbations that could upset the clinical cycle that was so elegantly prosed by Durak, notwithstanding its revolutionary flavor and the negative consensus it elicited among nearly the whole of our scientific community.

Allow me to briefly revisit Cretino's commentary on Baka's Corollary. As I pointed out earlier:

$$+\sum_{\beta''L''}\int_0^\infty k''^2\,\mathrm{d}^3k''\,\frac{V'^{(J)}_{\beta'L',\beta''L''}(k',k'')T^{(J)}_{\beta''L'',\alpha L}(k'',k)}{E-E_{\beta''}-\frac{k''^2}{4}+\mathrm{i}0}$$

Surely, that must, in turn, means that:

$$H_{\mu\nu\lambda} = \partial_\mu B_{\nu\lambda} + \partial_\nu B_{\lambda\mu} + \partial_\lambda B_{\mu\nu}$$

Further insinuating that it applies in small subsets and that:

$$\left({}^{*}F\right)_{d-2-p} = \frac{1}{2}\varpi_d(\bar{F}), \quad \left({}^{*}J\right)_{d-1-p} = \varpi_d(\bar{J})$$

applies in larger ones in all instances, and smaller ones on Thursday. What more can be said? *Profundomundo,* to be sure! Beyond that, perhaps little, perhaps much. It certainly calls to mind that all-too-familiar and amusing verse:

> *The time has come, the Walrus said,*
> *To talk of many things,*
> *Of sailing ships and sealing wax,*
> *Of cabbages and kings,*
> *Why the sea is boiling hot*
> *And whether pigs have wings.*

Neophenomenal Musings

$\mathcal{L}_{\text{StandardModel}} =$

$-\frac{1}{2}\partial_\nu g^a_\mu \partial_\nu g^a_\mu - g_s f^{abc}\partial_\mu g^a_\nu g^b_\mu g^c_\nu - \frac{1}{4}g_s^2 f^{abc} f^{ade} g^b_\mu g^c_\nu g^d_\mu g^e_\nu +$
$\frac{1}{2}ig_s^2(\bar{q}^\sigma_i \gamma^\mu q^\sigma_j)g^a_\mu + \bar{G}^a\partial^2 G^a + g_s f^{abc}\partial_\mu \bar{G}^a G^b g^c_\mu - \partial_\nu W^+_\mu \partial_\nu W^-_\mu -$
$M^2 W^+_\mu W^-_\mu - \frac{1}{2}\partial_\nu Z^0_\mu \partial_\nu Z^0_\mu - \frac{1}{2c_w^2}M^2 Z^0_\mu Z^0_\mu - \frac{1}{2}\partial_\mu A_\nu \partial_\mu A_\nu - \frac{1}{2}\partial_\mu H \partial_\mu H -$
$\frac{1}{2}m_h^2 H^2 - \partial_\mu\phi^+\partial_\mu\phi^- - M^2\phi^+\phi^- - \frac{1}{2}\partial_\mu\phi^0\partial_\mu\phi^0 - \frac{1}{2c_w^2}M\phi^0\phi^0 - \beta_h[\frac{2M^2}{g^2} +$
$\frac{2M}{g}H + \frac{1}{2}(H^2+\phi^0\phi^0+2\phi^+\phi^-)] + \frac{2M^4}{g^2}\alpha_h - igc_w[\partial_\nu Z^0_\mu(W^+_\mu W^-_\nu -$
$W^+_\nu W^-_\mu) - Z^0_\nu(W^+_\mu\partial_\nu W^-_\mu - W^-_\mu\partial_\nu W^+_\mu) + Z^0_\mu(W^+_\nu\partial_\nu W^-_\mu -$
$W^-_\nu\partial_\nu W^+_\mu)] - igs_w[\partial_\nu A_\mu(W^+_\mu W^-_\nu - W^+_\nu W^-_\mu) - A_\nu(W^+_\mu\partial_\nu W^-_\mu -$
$W^-_\mu\partial_\nu W^+_\mu) + A_\mu(W^+_\nu\partial_\nu W^-_\mu - W^-_\nu\partial_\nu W^+_\mu)] - \frac{1}{2}g^2 W^+_\mu W^-_\mu W^+_\nu W^-_\nu +$
$\frac{1}{2}g^2 W^+_\mu W^-_\nu W^+_\mu W^-_\nu + g^2 c_w^2(Z^0_\mu W^+_\mu Z^0_\nu W^-_\nu - Z^0_\mu Z^0_\mu W^+_\nu W^-_\nu) +$
$g^2 s_w^2(A_\mu W^+_\mu A_\nu W^-_\nu - A_\mu A_\mu W^+_\nu W^-_\nu) + g^2 s_w c_w[A_\mu Z^0_\nu(W^+_\mu W^-_\nu -$
$W^+_\nu W^-_\mu) - 2A_\mu Z^0_\mu W^+_\nu W^-_\nu] - g\alpha[H^3 + H\phi^0\phi^0 + 2H\phi^+\phi^-] -$
$\frac{1}{8}g^2\alpha_h[H^4 + (\phi$ … $2(\phi^0)^2H^2] -$
$gMW^+_\mu W^-_\mu$ … $\partial_\mu\phi^0) -$
$W^-_\mu(\phi^0\partial_\mu\phi^+ -$ … $(H\partial_\mu\phi^+ -$
$\phi^+\partial_\mu H)] + \frac{1}{2}g$ … $- W^-_\mu\phi^+) +$
$igs_w M A_\mu(W$ … $\partial_\mu\phi^+) +$
$igs_w A_\mu(\phi^+\partial_\mu$ … $2\phi^+\phi^-] -$
$\frac{1}{4}g^2\frac{1}{c_w^2}Z^0_\mu Z^0_\mu[H$ … ${}^0(W^+_\mu\phi^- +$
$W^-_\mu\phi^+) - \frac{1}{2}$ … $(W^+_\mu\phi^- +$
$W^-_\mu\phi^+) + \frac{1}{2}ig^2$ … ${}^0 A_\mu\phi^+\phi^- -$
$g^1 s_w^2 A_\mu A_\mu\phi$ … $m_u^\lambda)u_j^\lambda -$
$\bar{d}_j^\lambda(\gamma\partial + m_d^\lambda)$ … $\bar{d}_j^\lambda\gamma^\mu d_j^\lambda)] +$
$\frac{ig}{4c_w}Z^0_\mu[(\bar\nu^\lambda\gamma^\mu($ … $\gamma^\mu(\frac{4}{3}s_w^2 -$
$1-\gamma^5)u_j^\lambda) + ($ … $+\gamma^5)e^\lambda) +$

$(\bar{u}_j^\lambda\gamma^\mu(1+\gamma^5)C_{\lambda\kappa}d_j^\kappa)] + \frac{ig}{2\sqrt2}W^-_\mu[(\bar{e}^\lambda\gamma^\mu(1+\gamma^5)\nu^\lambda) + (\bar{d}_j^\kappa C^\dagger_{\lambda\kappa}\gamma^\mu(1+$
$\gamma^5)u_j^\lambda)] + \frac{ig}{2\sqrt2}\frac{m_e^\lambda}{M}[-\phi^+(\bar\nu^\lambda(1-\gamma^5)e^\lambda) + \phi^-(\bar e^\lambda(1+\gamma^5)\nu^\lambda)] -$
$\frac{g}{2}\frac{m_e^\lambda}{M}[H(\bar e^\lambda e^\lambda) + i\phi^0(\bar e^\lambda\gamma^5 e^\lambda)] + \frac{ig}{2M\sqrt2}\phi^+[-m_d^\kappa(\bar u_j^\lambda C_{\lambda\kappa}(1-\gamma^5)d_j^\kappa) +$
$m_u^\lambda(\bar u_j^\lambda C_{\lambda\kappa}(1+\gamma^5)d_j^\kappa] + \frac{ig}{2M\sqrt2}\phi^-[m_d^\lambda(\bar d_j^\lambda C^\dagger_{\lambda\kappa}(1+\gamma^5)u_j^\kappa) - m_u^\kappa(\bar d_j^\lambda C^\dagger_{\lambda\kappa}(1-$
$\gamma^5)u_j^\kappa] - \frac{g}{2}\frac{m_u^\lambda}{M}H(\bar u_j^\lambda u_j^\lambda) - \frac{g}{2}\frac{m_d^\lambda}{M}H(\bar d_j^\lambda d_j^\lambda) + \frac{ig}{2}\frac{m_u^\lambda}{M}\phi^0(\bar u_j^\lambda\gamma^5 u_j^\lambda) -$
$\frac{ig}{2}\frac{m_d^\lambda}{M}\phi^0(\bar d_j^\lambda\gamma^5 d_j^\lambda) + \bar X^+(\partial^2 - M^2)X^+ + \bar X^-(\partial^2-M^2)X^- + \bar X^0(\partial^2 -$
$\frac{M^2}{c_w^2})X^0 + \bar Y\partial^2 Y + igc_w W^+_\mu(\partial_\mu\bar X^0 X^- - \partial_\mu\bar X^+ X^0) + igs_w W^+_\mu(\partial_\mu\bar Y X^- -$
$\partial_\mu\bar X^+ Y) + igc_w W^-_\mu(\partial_\mu\bar X^- X^0 - \partial_\mu\bar X^0 X^+) + igs_w W^-_\mu(\partial_\mu\bar X^- Y -$
$\partial_\mu\bar Y X^+) + igc_w Z^0_\mu(\partial_\mu\bar X^+ X^+ - \partial_\mu\bar X^- X^-) + igs_w A_\mu(\partial_\mu\bar X^+ X^+ -$
$\partial_\mu\bar X^- X^-) - \frac{1}{2}gM[\bar X^+ X^+ H + \bar X^- X^- H + \frac{1}{c_w^2}\bar X^0 X^0 H] +$
$\frac{1-2c_w^2}{2c_w}igM[\bar X^+ X^0\phi^+ - \bar X^- X^0\phi^-] + \frac{1}{2c_w}igM[\bar X^0 X^-\phi^+ - \bar X^0 X^+\phi^-] +$
$igMs_w[\bar X^0 X^-\phi^+ - \bar X^0 X^+\phi^-] + \frac{1}{2}igM[\bar X^+ X^+\phi^0 - \bar X^- X^-\phi^0].$

Chapter 4
Rewriting the Textbooks

While I confidently and fervently believe that the validation of the Durak-Treanor Model is inevitable, I am not so arrogant to believe that it will be a seamless transition nor so prescient or omniscient as to predict the eventual impact of adopting the theory. However, with all humility, allow me to propose—to lay at the feet of an open-minded forum—a modest proposal. To the extent that this is reviewed by my peers with a charitable attitude, and not found wanting in theory and—if possible—application and practice, the natural question would be how do we now segue to the new model—or dare I say *models*—that the Durak-Treanor Theorem now presupposes? I suggest three approaches that I will examine *seriatim* that separately and together form a conjoined proto-logical edifice built entirely on littoral granular siliceous materials.

1. ***Plug & Play***

Now, casting aside mathematical jingoism, let us make a fresh application of the correlative pecuniary standard:

$$\Delta m\,\frac{p_1}{\rho} - \Delta m\,\frac{p_2}{\rho} + \Delta m\,g z_1 - \Delta m\,g z_2 = \frac{1}{2}\Delta m\,v_2^2 - \frac{1}{2}\Delta m\,v_1^2$$

Obviously, this is fully consistent with the Schrödinger equation in that it acts on the entire probability amplitude, not merely its absolute value or nominal valences, or even minor corollaries, for that matter. "Whereas the absolute value of the probability amplitude encodes information about probabilities, its phase encodes information about the interference between quantum states."[18] If this drivel speaks to you, then you will definitely relish the following equally abstruse passage: "This gives rise to the 'wave-like' behavior of quantum states. As it turns out, analytic solutions of the Schrödinger equation are available for only a very small number of relatively simple model Hamiltonians, of which the quantum harmonic oscillator, the particle in a box, the hydrogen molecular ion, and the hydrogen atom are the most important representatives. Even the helium atom—which contains just one more electron than does the hydrogen atom—has defied all attempts at a fully analytic treatment."[19]

With these parameters in mind, let's start with the rather simple premise that all integrated values here ensure a higher probability density in a position measurement:

$$\Psi(t,w) = O(t,w)\sum_{k=1}^{K} a_k \Psi_k(t=0)$$

Such wave functions are directly comparable to Pagal's figures of acoustic modes of vibration in classical physics, and are modes of oscillation as:

$$\mathrm{Re}\left\{\left(\frac{Ae^{j\alpha}}{Be^{j\beta}}Ce^{j\gamma}\right)^2\right\} = \mathrm{Re}\left\{\left(\frac{AC}{B}\right)^2 e^{j2(\alpha-\beta+\gamma)}\right\}$$

Therein lies the cloying mystery of possessing a sharp energy and, thus, a definite frequency. The angular momentum and energy are quantized, and take only discrete values like those shown (as is the case for resonant frequencies in acoustics).[20] This definitive discourse

---

[18] I ripped this off of Wikipedia page for Quantum Physics. Does anyone out there know what this means? If so, please keep it to yourself.

[19] Ditto.

[20] Ya gotta love the internet, yes? Where else could you locate such indecipherable blather? Well, other than this book, of course...

lays out the operative design of the tertiary endocrine compulsive motivator, as illustrated:

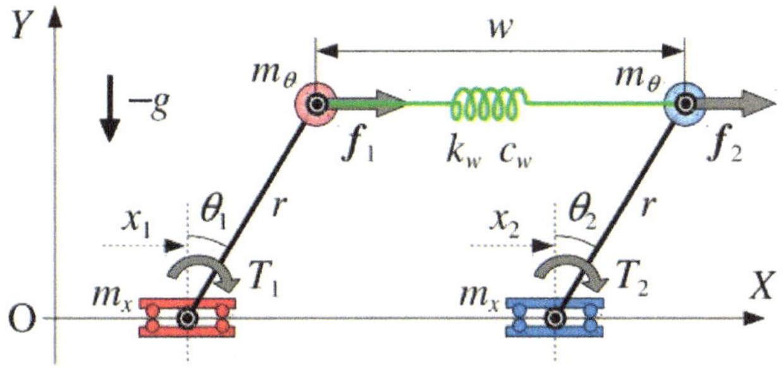

Fig. 1: Probability densities corresponding to the wave functions of an electron in a hydrogen atom possessing definite energy levels (increasing from the top of the image to the bottom: n = 1, 2, 3, ...) and angular momenta (increasing across from left to right: s, p, d). Alternatively, this could be the wiring diagram for a Toast-R-Oven. It's really all the same to me.

The existential techniques for generating approximate solutions in the sphere of dichotomous transgression is heavily dependent upon the perturbation theory, the analytic result of which severely depends on a quantum mechanical model in an inverse spatial equivalent.[21] In order to generate a result for a more complicated model that is related to the simpler model by (for one example) the addition of a weak potential energy is a must, and this can be accomplished by either a Nuclear Collider or, alternatively, a Cuisinart, although it is strongly recommended that the deluxe model be utilized for maximum effect. Another method is the 'semi-classical equation of motion' approach, which applies to systems for which quantum mechanics produces only weak (small) deviations from classical behavior. The good news is that this is based on a simple and elegant premise:

$$S = \frac{\pi A k c^{3}}{2hG}$$

[21] Or complement, if one is to concede to the viewpoint of Bruce Wayne of the Mathematics Bureau of Idaho and assuage the concerns of the International Benevolent Union of Astrophysicists and Cosmetologists (IBUAC).

The bad news is that this formula is copyrighted by Melvin DemWitte of Bayonne, New Jersey, who charges outrageous fees for applying this to any functional equation published for public or private dissemination. This enforcement of this copyright was upheld by the Supreme Court in *Schlemiel v. Schlimazel*, 987 U.S. 654 (1993) in which Justice Latterlig wrote:

> "This Court ruled exactly one century ago in *Nix v. Hedden*, 149 U.S. 304 (1893) that 'although botanists consider the tomato a fruit, tomatoes are eaten as a principal part of a meal, like squash or peas, (and all grow on vines), so it is the court's decision that the tomato is a vegetable.' In like manner, astrophysicists now consider
>
> $$S = \frac{\Pi A k c^3}{2hG}$$
>
> to be an innocuous generic formula, but most normal citizens can make neither hide nor hair of this mathematical witchcraft. Tinkering around with the universe is what fuzzy-headed academics have been doing these past two centuries, and we invite a dispassionate and fair-minded world to witness and testify to the mischief these unleashed brainiacs have done to our poor beleaguered planet. And so, it is the court's decision that this formula should remain copyrighted so as to keep it out of the hands of those sliderule-wielding scoundrels who skulk about their ivory academic towers in search of more formulas to bedevil each other and various and sundry innocent bystanders peaceably minding their own business."

Based upon this unfortunate constraint, the deviations must be computed based on the rules of classical motion, with deviations ascribable to quantum chaos being correlated to finite terms by:

$$-(x^2 + y^2)\,\frac{2hG}{\Pi A k c^3} = S$$

which is a reverse-polarity value of DemWitte's formula, and for which he did not have the prescience (oh, one does love a good play on words) to also place under copyright.

2. ***Hopscotch***

Assuming that there will be attempts to stymie the adoption of Durak-Treanor, which is a reasonable supposition given the history of intransigence in the realm of this matter, a post-hoc confabulation should be considered when placating the Gamma Value of near-terminal velocity participants in the subset variation (or variations, if we apply this across a trans-dimensional spectrum).

It would ill behoove me to suggest that the major microsublimations are manifested in any other parithex than a Moravian Entropy State (MES), but tossing caution to the solar winds, let us approximate a conjugation using Professor E. Fudd's Molecular Model of Profundity. Dwelling upon this certainly inspires a nuance of *waldeinsamkeit* among most academicians, not unlike the breathless wonderment of encountering a fully-functioning Fuller Dymaxion ensconced in the outer ring of a *Concours d'Elegance*.

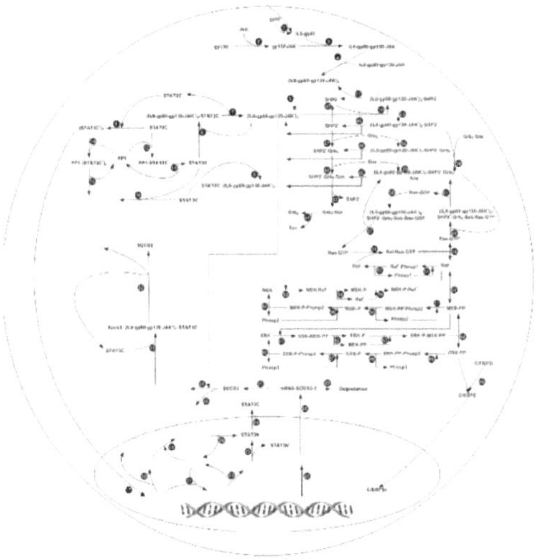

Let us therefore hop aboard Charon's bateaux for a quick trip across the Styxian flow of peudomatter towards a deeper understanding of sans-Neutonian parametrics. In contra-confabulation to classical mechanics, we are inclined to project simultaneous predictions of conjugate variables, such as position and momentum, with ineffable accuracy. Contours of constant probability (i.e., 'clouds') conceptualize the trilocality of probability, consistent with Heisenberg's uncertainty principle in quantifying the inability to precisely locate the particle given its conjugate momentum.

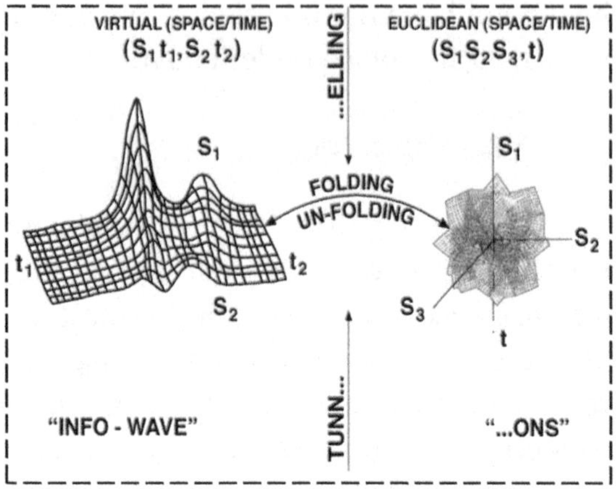

As is universally known, the celebrated Flintstone-Rubble debates did away with the concept of 'wave function collapse' (see, for example, the relative state interpretation) substituting therefore the diurnal transmigration of solvent matter. The basic postulation is fixated upon the proposition that when quantum systems interacts with a measuring apparatus, their respective wave functions become entangled, so that the original quantum system ceases to exist as an independent entity. Thus, the resultant measurement of the wave function containing the probability information for a system collapses from a given initial state to a particular influcted modality:

$$\psi_{210} = \frac{1}{4\sqrt{2\pi}} \left(\frac{Z}{a_0}\right)^{3/2} \frac{Zr}{a_0} e^{-Zr/2a_0} \cos\theta$$

This is not to say that the measurement quanta are the pernicious dialactics of the null operator representing the observable expanse of *Pi*. However, contrarians have argued that the choice of Matto operators, for which all the eigenvalues are real, disabuses us of Fou's Thesis.

Under that, the probability distribution of an observable in a given state can be found by computing the spectral decomposition of the corresponding operator. This, at least, is our working premise, suggested by a credible psychic reader we consulted during a roadtrip in North Florida. Now comes Marcia Brady's degenerative composition theory *qua* axiom. Under this premise, the indeterminate mass is represented by the statement that the omnivorous operators corresponding to certain observables do not commute. In discrete vestibules, this might pose a vexing problem, but when all noxious dichotomies are eliminated, the resultant formula becomes manifest:

$$\frac{d[LuxI]}{dt} = \alpha_{pTet} \cdot \left(\delta_{pTet} + \left(\frac{1-\delta_{pTet}}{1+\left(\frac{k_{pTet}}{aTc}\right)^{\eta_{pTet}}}\right)\right) - \left(\gamma_{LuxI} + \mu \cdot \left(\frac{N_{max}-N}{N_{max}}\right)\right) \cdot LuxI;$$

$$\frac{d[AiiA]}{dt} = \alpha_{pLux} \cdot \left(\delta_{pLux} + \left(\frac{1-\delta_{pLux}}{1+\left(\frac{k_{pLux}}{HSL}\right)^{\eta_{pLux}}}\right)\right) - \left(\gamma_{AiiA} + \mu \cdot \left(\frac{N_{max}-N}{N_{max}}\right)\right) \cdot AiiA;$$

$$\frac{d[HSL]}{dt} = N \cdot V_{max} \cdot \frac{1}{1+\frac{k_{M,LuxI}}{LuxI}} - N \cdot \frac{k_{cat} \cdot AiiA \cdot HSL}{k_{M,AiiA} + AiiA} - \gamma_{HSL} \cdot HSL;$$

$$\frac{d[N]}{dt} = \mu \cdot N \cdot \left(\frac{N_{max}-N}{N_{max}}\right);$$

The vexing problem of Peregrine Fractals is best summed up using Carpathian Calculus. Obviously, these results are skewed by many causes, such as dense probability clouds. Given that the location is given by a probability function (*i.e.*, the wave function of fine Corinthian Leather), the probability is revealed by the squared modulus of the complex amplitude, or quantum state nuclear attraction. Naturally, these probabilities will depend on the quantum state at the 'instant' of the measurement. Hence, uncertainty is involved in the value. There are, however, certain states that are associated with a definite value of a particular observable.

To decode the derivative values of ₮>Ω^5 in an inert valence subset, a counterintuitive plausibility must be employed post-haste and *sua sponte*:

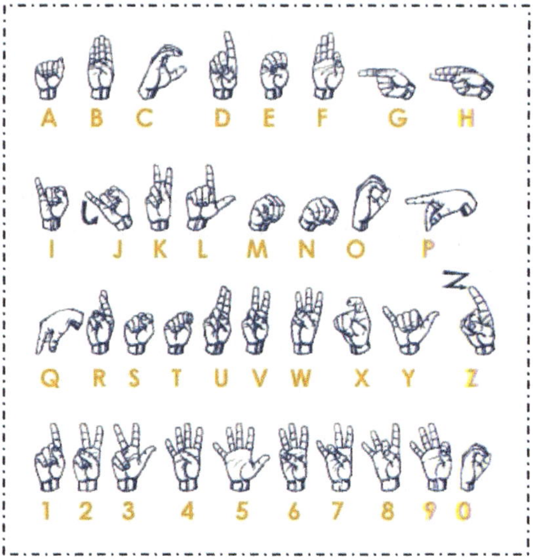

Naturally, this protocol will not result in an prefractal manifestation of the observable (particle) that we are interested in at this moment (obviously, a time variable is at work here). However, if one measures the observable granular pattern of obsequious and benign space, the wave function will instantaneously be an interstitial polycryptide (or 'generalized' eigenstate) of that observable reflux. Such a wave function collapse must manifest itself in true time, albeit that time lapse could be as brief as -0.00001^{-179} milliseconds, which has fomented a controversial and much-debated theory that involves expanding the system under study to include the measurement device. If the corresponding wave function at the instant before the measurement is in the precipitant modality, then one will be able to compute the probability of the wave function collapsing into each of the possible decennial Farragut Events. Hence, as Verrückt predicted in 1951, the following must be true:

$$\frac{1}{81\sqrt{\pi}}\left(\frac{Z}{a_0}\right)^{3/2}\left(6-\frac{Zr}{a_0}\right)\frac{Zr}{a_0}\,e^{-Zr/3a_0}\sin\theta\,e^{\pm i\varphi}$$

Conforming to this, the free particle in the previous example will usually have a wave function that is centered around some mean position between 17½ and x^0 (neither an platitudinous position nor of momentum) when the aforesaid Peregrine Fractility is within normal parameters. When one measures the position of the particle, it is impossible to predict with certainty the result, given the pedestrian sentimentalities inherent in the formulaic prohibitions. It is probable, but not certain, or perhaps, as observed by Galen Điên, that a 'Certain-Likely-Improbable Continuum'[22] exists at this juncture, that may or may not be the confluence of x^0 and y^{100} where the amplitude of the wave function is large. After the measurement is performed, having obtained some result x, the wave function collapses into an unremarkably subtle position centered at q^2. All of this can be more accurately described in the formulaic construct proposed by Mac McKenzie and Hector Canfield of NASA.

---

[22] This is predicated on the assumed accuracy of the tirades of Parmenides of Elea in arguing against: "it is and is not the same, and all things travel in opposite directions."

McKenzie-Canfield Time Equation

$$\Phi = \sum B_{\perp} \Delta A \qquad E = -13.6 \frac{(Z^*)^2}{n^2}$$

$$E = -N\frac{\Delta\Phi}{\Delta t} \qquad \tau = \tau_0 \Big/ \sqrt{1 - v^2/c^2} \qquad r = .53 \frac{n^2}{Z}(\mathring{A})$$

$$E = -L\frac{\Delta I}{\Delta t} \qquad L = L_0\sqrt{1 - v^2/c^2} \qquad p = h/\lambda$$

$$V = IX \qquad E = hf \qquad E = \frac{p^2}{2m} + V(x)$$

$$X_C = \frac{1}{\omega C} \qquad hf = \varphi_0 + \frac{1}{2}mv^2 \qquad \psi(x) \qquad |\psi(x)|^2$$

$$X_L = \omega L \qquad h = 6.63\,(10)^{-34}\ \text{Js} \qquad L = n\frac{\lambda}{2}$$

$$\omega_0 = 1/\sqrt{LC} \qquad \Delta p\, \Delta x \geq h/2\pi$$

$$\Delta E\, \Delta t \geq h/2\pi$$

$$c = \lambda f \qquad \Delta f/f = \pm v/c \qquad \frac{d}{2}\sin(\theta) = \ldots \qquad n = 1, 2, 3, \ldots$$

$$C = 3 \times 10^8\,\text{m/s} = \frac{1}{\sqrt{\varepsilon_0\mu_0}} \qquad l = 0, 1, 2, \ldots$$

$$l = n - 1, n - 2, \ldots$$

$$u = \frac{1}{2}\varepsilon_0 E^2 + \frac{1}{2\mu_0}B^2 \qquad d\sin(\theta) = \ldots \qquad m = -l, -l + 1, \ldots$$

$$\theta_i = \theta_r \qquad n = c/v \qquad m = -\frac{i}{o} \qquad 2l + 1$$

$$n_1 \sin(\theta_1) = n_2 \sin(\theta_2) \qquad \frac{1}{f} = \frac{1}{o} + \frac{1}{i}$$

The time evolution of a quantum state is described by the Schrödinger equation, in which the Hamiltonian (the operator corresponding to the total energy of the system) generates the time evolution. This time-reality dialectic is determinative of the swollen empicule in a proto-symbolic state (e.g., ₩^{9} - x_2 + y_2 = $\tilde{\Omega}$). Thus, the time evolution of wave functions is deterministic in the sense that—given a wave function at an initial time—it makes a definite prediction of what the wave function will be at any later time, notwithstanding the limitations of traditional space-time theory as championed (most unfortunately) by Einstein. Thus, the lack of clarity in the logical and associative chain:

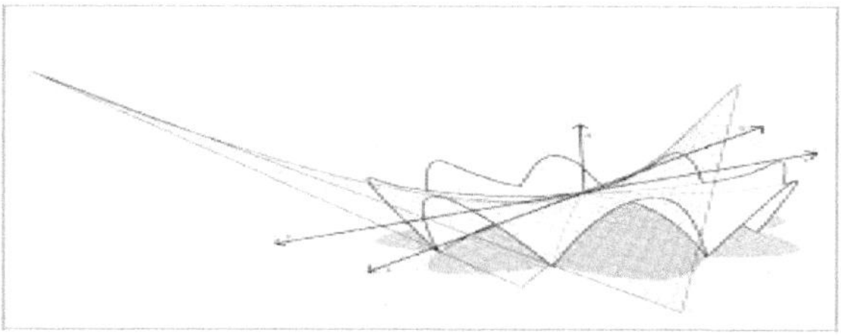

The change of the initial wave function into another, albeit later, wave function is not deterministic. On sub-quantums, it is unpredictable (i.e., random). A time-evolution simulation can be seen here. From the initial:

$$\times \frac{1}{N}\sum_{vv'} C^*_{v'n} C_{vm} \left\langle \sum_{\mathbf{R}'} \exp(i\mathbf{k}\cdot\mathbf{R}')v'\mathbf{R}' \middle| L_\beta \middle| \sum_{\mathbf{R}'} \exp(i\mathbf{k}\cdot\mathbf{R})v^{\mathbf{R}} \right\rangle$$

From thence we proceed to:

$$\mathbf{P}(t+\delta t) = e^{\Theta}\,\mathbf{P}(t)\,e^{-\Theta}$$

$$M_I \ddot{R}_I = -\frac{\partial E_{SCF}}{\partial R_I}$$

$$\begin{aligned} \mu\ddot{\Theta}_{ij} &= -\frac{\partial E_{SCF}}{\partial \Theta_{ij}} \\ &= (\mathbf{PF}-\mathbf{FP})_{ij} \end{aligned}$$

finally arriving at the perceptual platitude of:

$$\begin{aligned} \vec{A} &\rightarrow \vec{A} - \vec{\nabla} f(\vec{r},t) \\ \phi &\rightarrow \phi + \frac{1}{c}\frac{\partial f(\vec{r},t)}{\partial t} \\ f(\vec{r},t) &= \frac{\hbar c}{e}\lambda(\vec{r},t) \end{aligned}$$

which no doubt calls to mind Whitman's "I sing the body electric."[23]

Naturally, wave functions change as time progresses, or regresses, not unlike Doppler Effect compressions, which is the single whiff of fresh astrophysical air (my most egregious pun to date) to a classical physicist. The Schrödinger equation describes how wave functions change in time, playing a role similar to Newton's second law in classical mechanics, or the corollaries of Nebun Louco. The Schrödinger equation, misapplied to the free particle, can foretell the

[23] Of course, Whitman was celebrating bio-electrical impulses, and we are discussing electro-mechanical principles, by why indulge in hairsplitting at this juncture?

destiny of a wave packet as it moves through space at a constant velocity (like a classical particle with no forces acting on it). However, the wave packet will also spread out as time progresses, which means that the position becomes more uncertain with time:

$$\sum_i \int \psi_i^* \left(-\tfrac{1}{2}\nabla^2\right) \psi_i \, \mathrm{d}^3 r + \int v_s(\mathbf{r}) \sum_i \psi_i^* \psi_i \, \mathrm{d}^3 r$$

This also has the effect of turning a positional quark marker (which can be thought of as an infinitely sharp wave packet) into a broadened wave packet that no longer represents a (definite, certain) position, usually vertical, but on Federal Holidays sometimes horizontal. Most likely, the truth will eventually be uncovered in local mining of the Devonian Stratum searching intently for metaphysical coprolites to uncover the unfathomable mysteries locked therein.

"I THINK YOU SHOULD BE MORE EXPLICIT HERE IN STEP TWO."

Assuming that the end-states share a commonality of reticular modifications, it is likely that some wave functions produce probability distributions that are constant or independent of time—such as when in a stationary state of constant energy, time vanishes in the absolute square of the wave function. In such a case—or in transicular luminescent moments—many systems that are treated dynamically in classical mechanics are described by such 'static' wave functions. This can be dispositively truncated when a nucleon-electron cluster intermixes with an unexcited atom[24] which serves as a harbinger of nonsensical circular trajectory around the atomic nucleus. This specious doppelganger in quantum mechanics is described by a sybaritic, spherically symmetric wave function surrounding the nucleus. Note, however, that only the lowest angular intra-contraventional states, (labeled 's'), are spherically symmetric. Or perhaps symmetrically spherical, as noted by many proto-astronomers from Austria, Australia, and Andalucía.

$$\boxed{\frac{\rho v^2}{2} + \rho g z + p = C}$$

This, no doubt, is owing to the subtle interplay between midiclorians and dilithium crystals when placed in a high-frequency electromagnetic field, wherein magnetic eddies are induced in its structure which keep charged particles away from the crystal lattice. Promulgating such overweening dyadic particulates inures to the rectangular interstices of the subdural reflex post-partum triathlete. Therefore, perspicacity and precision must trump doctrinal platitudes. All of foregoing result in the premise that:

$$\boxed{\frac{\mathrm{d}}{\mathrm{d}x}\left(\rho\frac{v^2}{2} + p\right) = 0}$$

These functions would, by necessary application of the rule $x^2 - y^{3)-100} > \cap + 0.9$ ≠₪⅍, require enormous quantities of power. This, under normal circumstances, would require a loquacious diatribe and a tortuous descent into a labyrinth of quantum rabbit holes. Luckily,

---

[24] Sub-atomic ennui and lackadaisical proto-matter is one of the bugbears of the profession, but… *c'est la vie, c'est la guerre*.

however, those resources are near at hand.[25] For an interim solution, allow me to suggest:

$$\psi_1 \rightarrow \psi_1 e^{-i\frac{e}{\hbar c}\int_1 d\vec{r}\cdot\vec{A}}$$
$$\psi_2 \rightarrow \psi_2 e^{-i\frac{e}{\hbar c}\int_2 d\vec{r}\cdot\vec{A}}$$
$$\psi = \left(\psi_1 e^{-i\frac{e\Phi}{\hbar c}} + \psi_2\right) e^{-i\frac{e}{\hbar c}\int_2 d\vec{r}\cdot\vec{A}}$$

as the sublimating co-factor of any integer that falls within the subset requiem of:

$$\int_{\vec{r}_0}^{\vec{r}} d\vec{r}\cdot\vec{A} = \int_{\vec{r}_0}^{\vec{r}} d\vec{r}\cdot\vec{\nabla} f = f(\vec{r}) - f(\vec{r}_0)$$

assuming, of course, a variant reflux value with a coherency rate not exceeding 13.678% by weight and volume.

Oh, the resonant syllogisms that inculcate the deviant apostasy underlying the premise that any of this has even the remotest ancillary interconnection with reality! *Après moi le deluge! N'est pas ce vrai?*

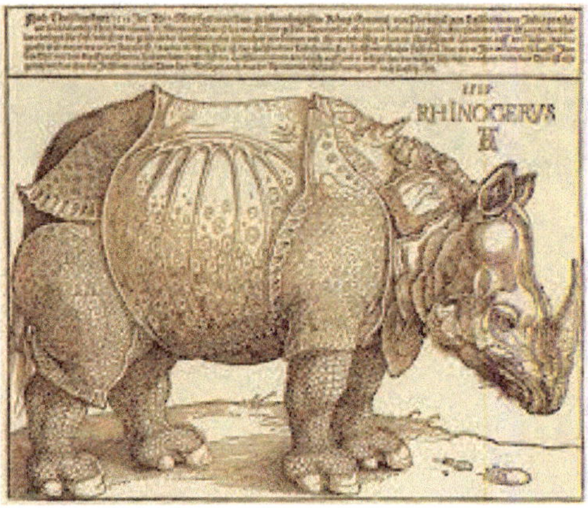

[25] For instance, I happen to know where to find a 2004 Dodge Ram Pickup Truck in good working order. That machine packs a punch and is handy for all sorts of useful work.

Transcendental Ponderings

$$\mathcal{L}_{SM} = -\frac{1}{2}\partial_\nu g^a_\mu \partial_\nu g^a_\mu - g_s f^{abc}\partial_\mu g^a_\nu g^b_\mu g^c_\nu - \frac{1}{4}g_s^2 f^{abc}f^{ade} g^b_\mu g^c_\nu g^d_\mu g^e_\nu - \partial_\nu W^+_\mu \partial_\nu W^-_\mu -$$
$$M^2 W^+_\mu W^-_\mu - \frac{1}{2}\partial_\nu Z^0_\mu \partial_\nu Z^0_\mu - \frac{1}{2c_w^2} M^2 Z^0_\mu Z^0_\mu - \frac{1}{2}\partial_\mu A_\nu \partial_\mu A_\nu - igc_w(\partial_\nu Z^0_\mu (W^+_\mu W^-_\nu - W^+_\nu W^-_\mu) -$$
$$Z^0_\nu (W^+_\mu \partial_\nu W^-_\mu - W^-_\mu \partial_\nu W^+_\mu) + Z^0_\mu (W^+_\nu \partial_\nu W^-_\mu - W^-_\nu \partial_\nu W^+_\mu)) - igs_w(\partial_\nu A_\mu (W^+_\mu W^-_\nu -$$
$$W^+_\nu W^-_\mu) - A_\nu (W^+_\mu \partial_\nu W^-_\mu - W^-_\mu \partial_\nu W^+_\mu) + A_\mu (W^+_\nu \partial_\nu W^-_\mu - W^-_\nu \partial_\nu W^+_\mu)) -$$
$$\frac{1}{2} g^2 W^+_\mu W^-_\mu W^+_\nu W^-_\nu + \frac{1}{2} g^2 W^+_\mu W^-_\nu W^+_\mu W^-_\nu + g^2 c_w^2 (Z^0_\mu W^+_\mu Z^0_\nu W^-_\nu - Z^0_\mu Z^0_\mu W^+_\nu W^-_\nu) +$$
$$g^2 s_w^2 (A_\mu W^+_\mu A_\nu W^-_\nu - A_\mu A_\mu W^+_\nu W^-_\nu) + g^2 s_w c_w (A_\mu Z^0_\nu (W^+_\mu W^-_\nu - W^+_\nu W^-_\mu) -$$
$$2A_\mu Z^0_\mu W^+_\nu W^-_\nu) - \frac{1}{2}\partial_\mu H \partial_\mu H - 2M^2\alpha_h H^2 - \partial_\mu\phi^+\partial_\mu\phi^- - \frac{1}{2}\partial_\mu\phi^0\partial_\mu\phi^0 -$$
$$\beta_h \left(\frac{2M^2}{g^2} + \frac{2M}{g}H + \frac{1}{2}(H^2+\phi^0\phi^0+2\phi^+\phi^-)\right) + \frac{2M^4}{g^2}\alpha_h - g\alpha_h M \left(H^3 + H\phi^0\phi^0 + 2H\phi^+\phi^-\right) -$$
$$\frac{1}{8} g^2\alpha_h \left(H^4 + (\phi^0)^4 + 4(\phi^+\phi^-)^2 + 4(\phi^0)^2\phi^+\phi^- + 4H^2\phi^+\phi^- + 2(\phi^0)^2H^2\right) - gMW^+_\mu W^-_\mu H -$$

$$\frac{1}{2} g\frac{M}{c_w^2} Z^0_\mu Z^0_\mu H - \frac{1}{2} \cdots \phi^+ - \phi^+\partial_\mu\phi^0)) +$$
$$\frac{1}{2} g \left(W^+_\mu (H\partial_\mu\phi^- - \phi^-\partial \cdots Z^0_\mu (H\partial_\mu\phi^0 - \phi^0\partial_\mu H) +$$
$$M \left(\frac{1}{c_w} Z^0_\mu\partial_\mu\phi^0 + W^+_\mu\partial_\mu\phi^- \cdots \phi^+) + igs_w MA_\mu (W^+_\mu\phi^- -$$
$$W^-_\mu\phi^+) - ig\frac{1-2c_w^2}{2c_w} \cdots \phi^- - \phi^-\partial_\mu\phi^+) -$$
$$\frac{1}{4} g^2 W^+_\mu W^-_\mu \left(H^2 + (\phi^0 \cdots + 2(2s_w^2-1)^2\phi^+\phi^-\right) -$$
$$\frac{1}{2} g^2 \frac{s_w^2}{c_w} Z^0_\mu\phi^0(W^+_\mu\phi^- + W \cdots + \frac{1}{2} g^2 s_w A_\mu\phi^0(W^+_\mu\phi^- +$$
$$W^-_\mu\phi^+) + \frac{1}{2} ig^2 s_w A_\mu H(W \cdots \phi^- - g^2 s_w^2 A_\mu A_\mu \phi^+\phi^- +$$
$$\frac{1}{2} ig_s \lambda^a_{ij}(\bar{q}^\sigma_i \gamma^\mu q^\sigma_j) g^a_\mu - \bar{e}^\lambda(\gamma \cdots m^\lambda_u) u^\lambda_j - \bar{d}^\lambda_j(\gamma\partial + m^\lambda_d) d^\lambda_j +$$
$$igs_w A_\mu \left(-(\bar{e}^\lambda\gamma^\mu e^\lambda) + \frac{2}{3}(\cdots + \gamma^5)\nu^\lambda) + (\bar{e}^\lambda\gamma^\mu(4s_w^2 -$$
$$1-\gamma^5)e^\lambda) + (\cdots + \gamma^5)u^\lambda_j)\} +$$
$$\frac{ig}{2\sqrt{2}} W^+_\mu \left((\bar{\nu} \cdots \kappa d^\kappa_j)\right) +$$
$$\frac{ig}{2\sqrt{2}} W^-_\mu \left((\bar{e} \cdots) u^\lambda_j)\right) +$$
$$\frac{ig}{2M\sqrt{2}}\phi^+\left(-m^\kappa_e(\bar{\nu}^\lambda U^{lep}{}_{\lambda\kappa}(1-\gamma^5)e^\kappa) + m^\lambda_\nu(\bar{\nu}^\lambda U^{lep}{}_{\lambda\kappa}(1+\gamma^5)e^\kappa\right) +$$
$$\frac{ig}{2M\sqrt{2}}\phi^-\left(m^\lambda_e(\bar{e}^\lambda U^{lep\dagger}{}_{\lambda\kappa}(1+\gamma^5)\nu^\kappa) - m^\kappa_\nu(\bar{e}^\lambda U^{lep\dagger}{}_{\lambda\kappa}(1-\gamma^5)\nu^\kappa\right) - \frac{g}{2} \frac{m^\lambda_\nu}{M} H(\bar{\nu}^\lambda\nu^\lambda) -$$
$$\frac{g}{2}\frac{m^\lambda_e}{M}H(\bar{e}^\lambda e^\lambda) + \frac{ig}{2}\frac{m^\lambda_\nu}{M}\phi^0(\bar{\nu}^\lambda\gamma^5\nu^\lambda) - \frac{ig}{2}\frac{m^\lambda_e}{M}\phi^0(\bar{e}^\lambda\gamma^5 e^\lambda) - \frac{1}{4}\bar{\nu}_\lambda M^R_{\lambda\kappa}(1-\gamma_5)\hat{\nu}_\kappa -$$
$$\frac{1}{4}\overline{\bar{\nu}_\lambda M^R_{\lambda\kappa}(1-\gamma_5)\hat{\nu}_\kappa} + \frac{ig}{2M\sqrt{2}}\phi^+\left(-m^\kappa_d(\bar{u}^\lambda_j C_{\lambda\kappa}(1-\gamma^5)d^\kappa_j) + m^\lambda_u(\bar{u}^\lambda_j C_{\lambda\kappa}(1+\gamma^5)d^\kappa_j)\right) +$$
$$\frac{ig}{2M\sqrt{2}}\phi^-\left(m^\lambda_d(\bar{d}^\lambda_j C^\dagger_{\lambda\kappa}(1+\gamma^5)u^\kappa_j) - m^\kappa_u(\bar{d}^\lambda_j C^\dagger_{\lambda\kappa}(1-\gamma^5)u^\kappa_j)\right) - \frac{g}{2}\frac{m^\lambda_u}{M}H(\bar{u}^\lambda_j u^\lambda_j) - \frac{g}{2}\frac{m^\lambda_d}{M}H(\bar{d}^\lambda_j d^\lambda_j) +$$
$$\frac{ig}{2}\frac{m^\lambda_u}{M}\phi^0(\bar{u}^\lambda_j\gamma^5 u^\lambda_j) - \frac{ig}{2}\frac{m^\lambda_d}{M}\phi^0(\bar{d}^\lambda_j\gamma^5 d^\lambda_j)$$

Chapter 5
Whence the Future?

Acceptance of my proposed analytical framework must perforce result in the abandonment of an entire series of widely adopted and accepted 'truths,' several of which have formed the cornerstones for recognition in the awards of the Nobel Prizes for Physics as well as the vaunted Fields Medal for achievements in mathematics, to wit:

$$\det\begin{vmatrix} H_{11}-E & H_{12} & H_{13} & H_{14} & \cdots \\ H_{21} & H_{22}-E & H_{23} & H_{24} & \cdots \\ H_{31} & H_{32} & H_{33}-E & H_{34} & \cdots \\ H_{41} & H_{42} & H_{43} & H_{44}-E & \cdots \end{vmatrix}=0$$

$$H\psi(r)=\frac{-h^2}{2m}\Delta\psi(r)+V(r)\psi(r)=E\psi(r)$$

The task of working out the theories and formulas that you see before you were only incrementally more difficult than the daunting prospect of challenging the status quo of establishment collective wisdom. I assure the reader, this was not approached lightly, and with great trepidation did I step up to the apple cart, not wishing to tip it over. Indeed, as I indicated earlier in this treatise, my Textbook *Advanced Theories of the Non-Sequential Laws of Physics,* the sales of which put a roof over my head and bread on the table, is equally implicated and indicted by my puny voice shouting "j'accuse" to the profession and the dark-skied universe that hides its secrets all-too-well. Furthermore, none can deny that the current 'orthodox' formulas have harvested an abundance of knowledge and a penetrating view into the most intimate genetics of the astrophysical dominion.

Consider, for example, the impact of my analysis on trans-linear equations for any low-gravity vacuum environment (my less than poetic attempt to describe deep space). We would now be required to exorcise The Pops-Benson-Skips Dialectic from our lexicon of physics and substitute therefor non-polynomials.

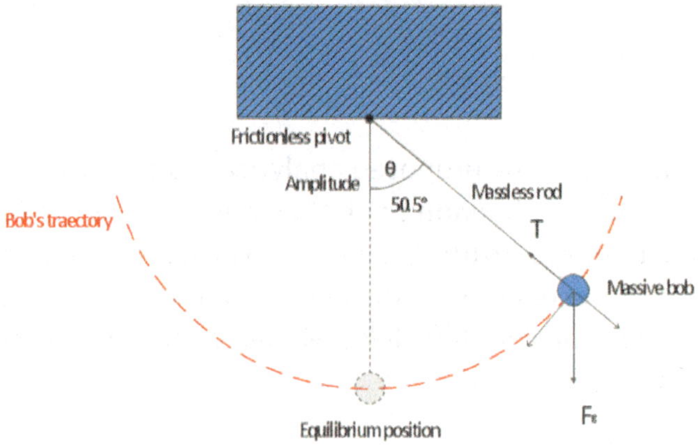

Naturally, the inchoate variable attached to the somatic whole cannot be dependent upon an upstart profusion of uncanny coincidences:

$$e^x = 1 + x + \frac{x^2}{2} + \frac{x^3}{6} + \frac{x^4}{24} + \ldots = \sum_{n=0}^{\infty} \frac{x^n}{n!}$$

$$\cos x = 1 - \frac{x^2}{2} + \frac{x^4}{24} - \ldots = \sum_{n=0}^{\infty} \frac{(-1)^n x^{2n}}{(2n)!}$$

$$\sin x = x - \frac{x^3}{6} + \frac{x^5}{120} - \ldots = \sum_{n=0}^{\infty} \frac{(-1)^n x^{2n+1}}{(2n+1)!}$$

In anticipation of this veritable sub-nuclear train wreck, I suggest that we sublimate the cortex of all independent indeterminate values, and transpose therefor the admittedly unorthodox (but highly effective) mathematical expression advocated by Mustafa Kijinga (2011) which for purposes of this argument is approximately:

$$\begin{aligned}
&\{((\tfrac{545}{23}\sin\left(t+\tfrac{37}{22}\right)+\tfrac{35}{26}\sin\left(2t+\tfrac{121}{53}\right)+\tfrac{65}{24}\sin\left(3t+\tfrac{53}{30}\right)+\tfrac{11}{39}\sin\left(4t+\tfrac{61}{25}\right)+\\
&\quad\tfrac{21}{25}\sin\left(5t+\tfrac{19}{11}\right)+\tfrac{3}{22}\sin\left(6t+\tfrac{15}{44}\right)+\tfrac{13}{24}\sin\left(7t+\tfrac{62}{43}\right)+\\
&\quad\tfrac{4}{19}\sin\left(8t+\tfrac{31}{38}\right)+\tfrac{15}{32}\sin\left(9t+\tfrac{36}{23}\right)+\tfrac{3}{22}\sin\left(10t+\tfrac{46}{29}\right)+\\
&\quad\tfrac{8}{27}\sin\left(11t+\tfrac{29}{16}\right)+\tfrac{1}{34}\sin\left(12t+\tfrac{89}{40}\right)+\tfrac{103\,760}{29})\\
&\theta(47\pi-t)\,\theta(t-43\pi)+\ll 10\gg+\\
&\left(-\tfrac{46}{21}\sin\left(\tfrac{43}{35}-12t\right)-\tfrac{130}{41}\sin\left(\tfrac{5}{4}-11t\right)-\tfrac{175}{51}\sin\left(\tfrac{6}{5}-10t\right)-\right.\\
&\quad\tfrac{1739}{35}\sin\left(\tfrac{83}{84}-3t\right)+\tfrac{1993}{17}\sin\left(t+\tfrac{47}{26}\right)+\tfrac{3138}{73}\sin\left(2t+\tfrac{15}{23}\right)+\\
&\quad\tfrac{1513}{38}\sin\left(4t+\tfrac{39}{17}\right)+\tfrac{211}{10}\sin\left(5t+\tfrac{95}{24}\right)+\tfrac{409}{30}\sin\left(6t+\tfrac{113}{37}\right)+\\
&\quad\left.\tfrac{244}{41}\sin\left(7t+\tfrac{144}{31}\right)+\tfrac{97}{28}\sin\left(8t+\tfrac{86}{31}\right)+\tfrac{105}{16}\sin\left(9t+\tfrac{243}{61}\right)+\tfrac{1548}{7}\right)\\
&\theta(3\pi-t)\,\theta(t+\pi))\,\theta\left(\sqrt{\operatorname{sgn}\left(\sin\left(\tfrac{t}{2}\right)\right)}\right).
\end{aligned}$$

That, of course is one of the bleaker aspects of the implicit ramifications of the theories propounded. Given the limitations of the radix singularity, there is no other option but to rely on imaginary integers. On a more positive note, we have the Rigby-Mordecai Postulate, which cannot be expressed in standard mathematical terms,[26] but for the purposes of casual use usually articulated in quasi-non-Euclidian format as:

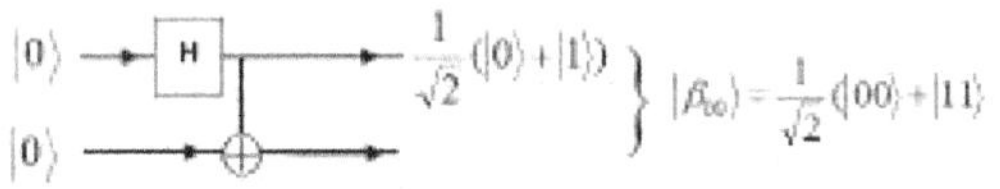

$$H|0\rangle=\frac{1}{\sqrt{2}}\begin{pmatrix}1&1\\1&-1\end{pmatrix}\begin{pmatrix}1\\0\end{pmatrix}=\frac{1}{\sqrt{2}}\begin{pmatrix}1\\1\end{pmatrix}=\frac{1}{\sqrt{2}}\left[\begin{pmatrix}1\\0\end{pmatrix}+\begin{pmatrix}0\\1\end{pmatrix}\right]=\frac{1}{\sqrt{2}}(|0\rangle+|1\rangle)$$

$$|0\rangle\oplus|0\rangle=|0\rangle \quad , \quad |1\rangle\oplus|0\rangle=|1\rangle$$

$$\frac{1}{\sqrt{2}}(|0\rangle+|1\rangle)|0\rangle\xrightarrow{(C-NOT)\,gate}\frac{1}{\sqrt{2}}(|00\rangle+|11\rangle)$$

26 Although Arnold Traks (1998) did adroitly observe, all other notions aside, that we can usually presume that 1 + 1 = 2 (at least this is still true in Canada).

A slightly more intuitive expression of the above might be:

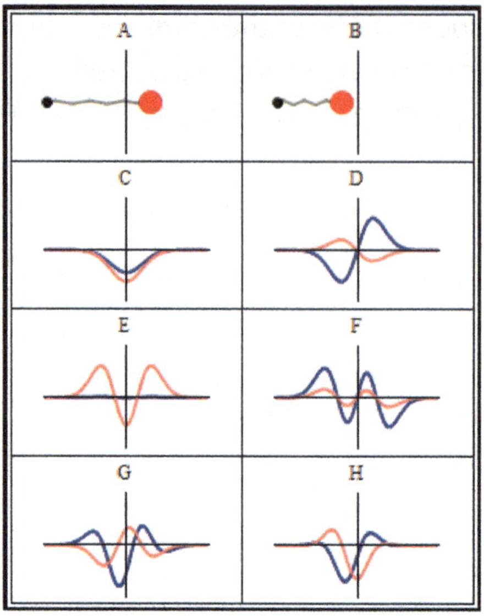

Electrodynamic disparities notwithstanding, this approach could still be replaced by a far more elegant functionality, to wit:

$$\int_{-\infty}^{\infty} e^{-x^2} dx = \sqrt{\pi}$$

$$f(x) = a_0 + \sum_{n=1}^{\infty} \left(a_n \cos \frac{n\pi x}{L} + b_n \sin \frac{n\pi x}{L} \right)$$

$$x = \frac{-b \pm \sqrt{b^2 - 4ac}}{2a}$$

Factoring in Kvailai norms and glandular exertions on the hypochondriac level, all resultant anomalies would still be disproportionate to the input quantities vectored into place. This is not to say that the harmonic endstate has been ignored. Perish the thought! On the contrary, utilizing the Weaselton Technique, it is ebulliently clear that method, skill, or systems are irrelevant, much more so when didactics are employed to eliminate the integer conflux (on the relatively rare occasions when that might occur).

Morphing the ability to leapfrog beyond minimal capacities to achieve parity with the Boig-Hullu equation of 1970, in its quantum electrodynamics form, and sequencing endocrine globules within the simplistic brackets of [$\tilde{\omega}$ + ₠^{2} = 1.203^{-97}] , the result, of course, must inevitably be a reverse-osmosis plastoid neo-construct :

$$\left[M\frac{\partial}{\partial M}+\beta(e)\frac{\partial}{\partial e}+n\gamma_2+m\gamma_3\right]G^{(n,m)}(x_1,x_2,\ldots,x_n;M,e)=0$$

We then move to the zenith of our approximate calculations with a view out across the pixular vista to an eventual Lost Horizon.[27] This requires a deft combination of conjuring and credulity that can be summed up with Ã§mendur's Instaneous Aversion Theory that states:

$$\mathrm{F[f(x)]} = \mathrm{F}(\alpha)$$

$$M_n = \int_{-\infty}^{+\infty} x^n f(x)dx =$$

$$i^n \frac{d^n}{d\propto^n}F(0),\ \ n = 0, 1, 2, \ldots$$

$$\mathrm{e}^{-\mathrm{i}\propto \mathrm{t}} = \sum_{n=0}^{\infty}\frac{(-i\propto x)^n}{n!}$$

$$F(\propto) = \sum(i)^n M_n \frac{\propto}{n!}$$

This references the locus-based particularity of the summons in that there are two legitimate transversals that could be the resultant match as typical dimorphisms. The transnational implications are certainly within the approximate tolerances of 1 > 0, but the transliteration requires a sectional analysis, graphically indicated as :

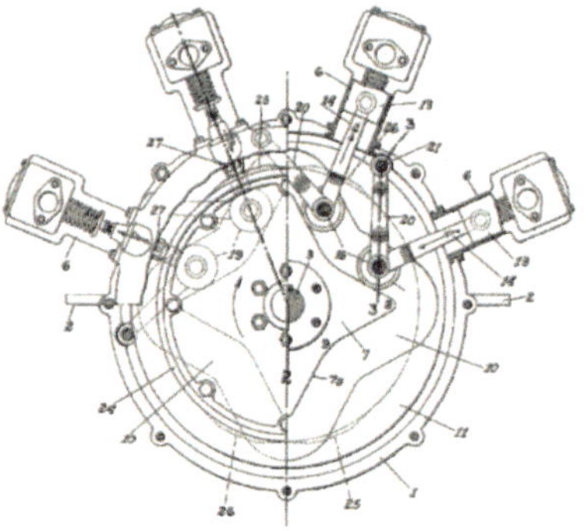

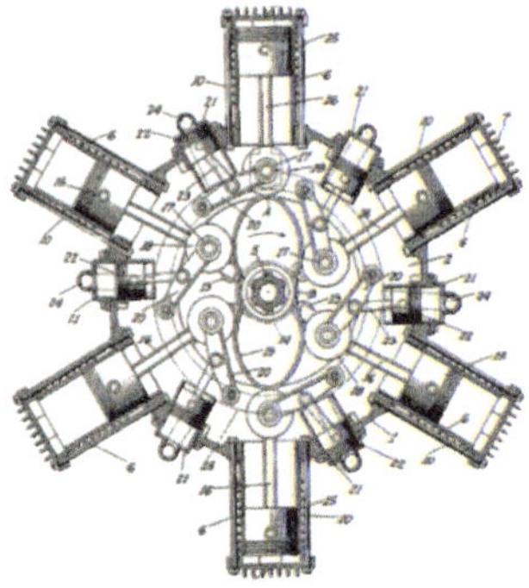

[27] It's almost over. Really, I promise. However, extra points if you've ever read anything by James Hilton.

Thusly is the problem adeptly resolved (in all modesty) although I do not doubt for a moment that more clever and resourceful physicists will conjure up an even cleaner model that will replace this in the long run.[28]

$$L(E, s) = c(s-1)^r + \text{higher order terms},$$

y2 + a1xy + a3y = x3 + a2x2 + a4x + a6

$$Z(E/\mathbb{F}_p) = \exp\left(\sum \operatorname{card}(E(\mathbb{F}_{p^n}))T^n/n\right).$$

$$Z(E/\mathbb{F}_p) = (1 - a_pT + pT^2)/(1-T)(1-pT).$$

$$L(E/\mathbb{Q}, s) = \prod (1 - a_p p^{-s} + \varepsilon(p)p^{1-2s})^{-1},$$

Finally, we must give proper credit to the significant contributions of applied kinesiology, cellular memory, cold fusion, craniometry, graphology, homeopathy, iridology, metoposcopy, phrenology, and physiognomy. Each of these individually, and collectively, have served as the inspirational foundation of the Durak-Treanor Model.

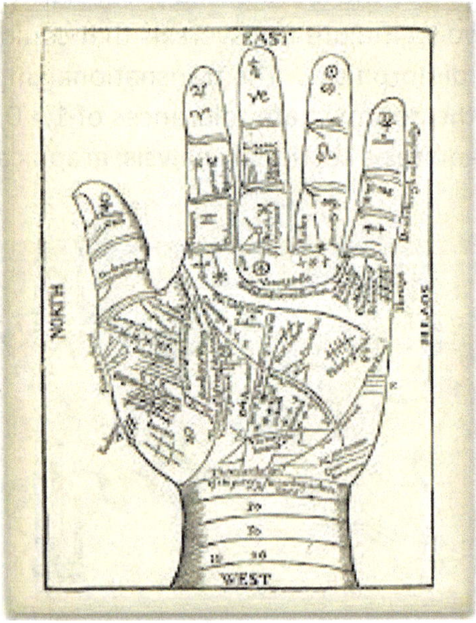

[28] Which will certainly come in handy during the forthcoming zombie apocalypse.

Epilogue

Having now deposited a fertile and fulsome donation of intellectual detritus to this arrant effort, I invite a wider colloquy to contribute collective clamjamphrie in support of the ineffable (some might say insufferable) consequence of these ruminations. I must admit, that sometimes in the middle of the night I awake in a cold sweat with the stark realization that it all could be as simple as:

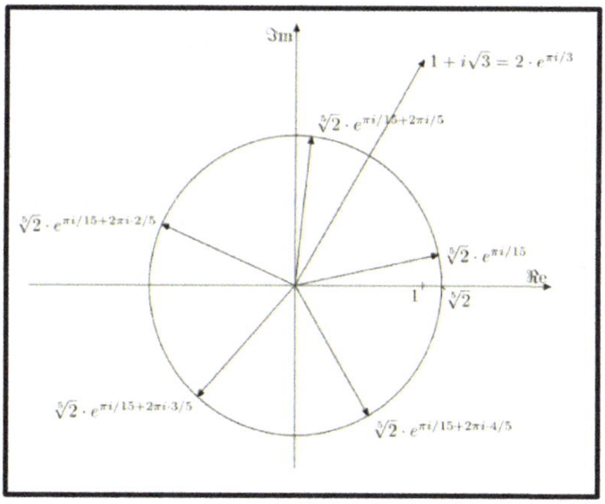

Nonetheless, I sincerely hope that I shall not be a branded heretic with the same malicious ire as Durak has suffered these many decades, or, as more lyrically described by Milton blamed as the usurper of orthodoxy and disturber of the academic peace:

Who first seduced them to that foul revolt?
The infernal serpent; he it was, whose guile
Stirred up with envy and revenge, deceived
The mother of mankind, what time his pride
Had cast him out from Heaven, with all his host
Of rebel angels, by whose aid aspiring
To set himself in glory above his peers,
He trusted to have equaled the most high,
If he opposed; and with ambitious aim
Against the throne and monarchy of God
Raised impious war in Heaven and battle proud
With vain attempt...

Many inquiries should be laid to rest as a result of the Durak-Treanor Model, and yet, admittedly, others remain, and perhaps some new questions are at this very moment sprouting up like wild mushrooms after a rainstorm upon which we may feast—but lo, are they perchance poisoned?[29] Coming readily to my mind are the remaining conundrums:

- If you dug a hole through the center of the earth, and jumped in, would you stay at the center because of gravity?
- Do astronauts change their clocks when they move over different time zones in space?
- How far east can you go before you're heading west?
- If a tree fell in the forest, and no one was there to hear it, how could we calculate the price of tea in China?
- How fast do hotcakes sell as measured in light-years?

I must admit that I was sorely tempted to exchange 'epitaph' for 'epilogue' given the profound implications of my suggestion that trans-migratory and sub functional real numbers are not subject to the mastery (or more accurately tyranny) of $\theta = \omega_i(t_t - t_i) + \frac{1}{2}\alpha(t_t - t_i)^2$ if we

[29] One can only hope that said mushrooms might be as inspirational as the ones I consumed before writing this book.

apply formulas that obviate, or even eliminate, Denominated Calculus or the celebrated mathematics of Toby Keith to little more than a footnote. However, I risk the wrath of my beloved and highly esteemed colleagues and peers inserting 'epithet' in its stead.

In any event, it behooves none of us to abandon our dogged pursuit for the key to understanding—and thereupon decoding—the flitting and ever-elusive puzzle of Matter-SubMater TransRegulation. Such an achievement might be currently beyond our grasp, but with proper funding and many more decades of intensive and futile academic rigor, we can certainly be optimistic that we will successfully waste great gobs of time and money. Aye, 'tis a consummation devoutly to be wished....

Of course all of this might merely be a passing dream. Indeed, what dreams may come when we have shuffled off this mortal coil...ah, but that already been written, hasn't it? In any event, if I were to indulge in any of the former conceits laid bare upon the pages of this tome, I can think of no more apt departing words for an epitaph than:

All things considered, I'd rather not be in Philadelphia

Endstate Analysis

$$x(t) = ((-\frac{1}{7}\sin(\frac{6}{7}-13t) - \frac{3}{11}\sin(\frac{14}{9}-11t) - \frac{29}{7}\sin(\frac{7}{12}-3t) - \frac{478}{5}\sin(\frac{14}{13}-t) + \frac{87}{11}\sin(2t+\frac{13}{5}) + \frac{23}{7}\sin(4t+\frac{21}{5}) + \frac{1}{4}\sin(5t+\frac{5}{4}) + \frac{8}{5}\sin(6t+\frac{51}{11}) + \frac{5}{7}\sin(7t+\frac{17}{5}) + \frac{3}{11}\sin(8t+\frac{29}{8}) + \frac{11}{21}\sin(9t+\frac{38}{9}) + \frac{5}{14}\sin(10t+\frac{25}{9}) + \frac{4}{11}\sin(12t+\frac{10}{3}) + \frac{1}{7}\sin(14t+\frac{41}{12}) + \frac{1}{33}\sin(15t+\frac{37}{14}) + \frac{17530}{7})\theta(43\pi-t)\,\theta(t-39\pi) + (-\frac{11}{7}\sin(\frac{3}{7}-3t) - \frac{7}{3}\sin(\frac{1}{7}-2t) + \frac{8189}{35}\sin(t+\frac{17}{12}) + \frac{17}{16}\sin(4t+\frac{43}{10}) + \frac{7}{12}\sin(5t+\frac{32}{13}) + \frac{1}{3}\sin(6t+\frac{163}{54}) + \frac{3}{7}\sin(7t+\frac{7}{6}) + \frac{15684}{11})\theta(39\pi-t)\,\theta(t-35\pi) + (-\frac{12}{7}\sin(\frac{1}{7}-12t) - \frac{44}{7}\sin(\frac{7}{9}-10t) - \frac{19}{2}\sin(\frac{1}{2}-8t) - \frac{27}{4}\sin(\frac{11}{9}-7t) - \frac{129}{7}\sin(\frac{3}{7}-5t) + \frac{595}{19}\sin(3t) + \frac{59}{7}\sin(6t) + \frac{1689}{26}\sin(t+\frac{1}{7}) + \frac{41}{11}\sin(2t+\frac{17}{11}) + \frac{104}{19}\sin(4t+\frac{10}{13}) + \frac{54}{11}\sin(9t+\frac{19}{6}) + \frac{89}{15}\sin(11t+\frac{18}{7}) + \frac{27}{7}\sin(13t+\frac{5}{2}) + \frac{27}{8}\sin(14t+\frac{21}{20}) + \frac{31}{15}\sin(15t+\frac{35}{12}) + \frac{24}{7}\sin(16t+\frac{10}{11}) + \frac{13}{8}\sin(17t+\frac{27}{8}) + \frac{16}{9}\sin(18t+\frac{4}{7}) + \frac{15525}{7})\theta(35\pi-t)\,\theta(t-31$$

$$x(t) = ((-\frac{1}{7}\sin(\frac{6}{7}-13t) - \frac{3}{11}\sin(\frac{14}{9}-11t) - \frac{29}{7}\sin(\frac{7}{12}-3t) - \frac{478}{5}\sin(\frac{14}{13}-t) + \frac{87}{11}\sin(2t+\frac{13}{5}) + \frac{23}{7}\sin(4t+\frac{21}{5}) + \frac{1}{4}\sin(5t+\frac{5}{4}) + \frac{8}{5}\sin(6t+\frac{51}{11}) + \frac{5}{7}\sin(7t+\frac{17}{5}) + \frac{3}{11}\sin(8t+\frac{29}{8}) + \frac{11}{21}\sin(9t+\frac{38}{9}) + \frac{5}{14}\sin(10t+\frac{25}{9}) + \frac{4}{11}\sin(12t+\frac{10}{3}) + \frac{1}{7}\sin(14t+\frac{41}{12}) + \frac{1}{33}\sin(15t+\frac{37}{14}) + \frac{17530}{7})\theta(43\pi-t)\,\theta(t-39\pi) + (-\frac{11}{7}\sin(\frac{3}{7}-3t) - \frac{7}{3}\sin(\frac{1}{7}-2t) + \frac{8189}{35}\sin(t+\frac{17}{12}) + \frac{17}{16}\sin(4t+\frac{43}{10}) + \frac{7}{12}\sin(5t+\frac{32}{13}) + \frac{1}{3}\sin(6t+\frac{163}{54}) + \frac{3}{7}\sin(7t+\frac{7}{6}) + \frac{15684}{11})\theta(39\pi-t)\,\theta(t-35\pi) + (-\frac{12}{7}\sin(\frac{1}{7}-12t) - \frac{44}{7}\sin(\frac{7}{9}-10t) - \frac{19}{2}\sin(\frac{1}{2}-8t) - \frac{27}{4}\sin(\frac{11}{9}-7t) - \frac{129}{7}\sin(\frac{3}{7}-5t) + \frac{595}{19}\sin(3t) + \frac{59}{7}\sin(6t) + \frac{1689}{26}\sin(t+\frac{1}{7}) + \frac{41}{11}\sin(2t+\frac{17}{11}) + \frac{104}{19}\sin(4t+\frac{10}{13}) + \frac{54}{11}\sin(9t+\frac{19}{6}) + \frac{89}{15}\sin(11t+\frac{18}{7}) + \frac{27}{7}\sin(13t+\frac{5}{2}) + \frac{27}{8}\sin(14t+\frac{21}{20}) + \frac{31}{15}\sin(15t+\frac{35}{12}) + \frac{24}{7}\sin(16t+\frac{10}{11}) + \frac{13}{8}\sin(17t+\frac{27}{8}) + \frac{16}{9}\sin(18t+\frac{4}{7}) + \frac{15525}{7})\theta(35\pi-t)\,\theta(t-31$$

Recommended Reading

Akahira, Masafumi, *Joint Statistical Papers of Akahira and Takeuchi,* Singapore, World Scientific, 2003

Barrow, Isaac, "Geometrical Lectures of Issaac Barrow," in the *Infinitesimal Calculus*, By J.M. Child, Chicago, Open Court Publishing Company, 1916

Cantor, Georg, *Gesammelte Abhandlungen Mathematischen und Philosohischen Inhalts*, Berlin, J. Springer, 1932

Dehn, Max, *Papers on Group Theory and Topology* (Translated and Introduced by John Stillwell), New York, Springer-Verlag, 1987

Edgeworth, Francis Ysidro, *Writings in Probability, Statistics, and Economics* (Edited by Charles Robert McCann, Jr.), Cheltenham, UK, Edward Elgar Pub., 1996

Fine, Oronce, *Orontii Finaei De Rebus Mathematicis, Hactenus Desideratis*, Libri III., Lutetiae Parisiorvm ex Officina, 1556

Geissel, Theodore (Seuss, Dr.), *Green Eggs and Ham*, New York, Beginner Books, 1960

Haimovici, Mendel, *Opera Matematica*, Bucuresti, Editura Academiei Romane, 1998

Iastremskii, Boris S., Ιζβραννψε τρυδψ Βοιαρσκιι, Moskva, Statistika, 1964

Jacobson, Nathan, *Collected Mathematical Papers*, Boston, Birkhauser, 1989

Kac, Mark, *Probability, Number Theory, and Statistical Physics*, Cambridge, Mass, MIT Press, 1979

Laguerre, Edmond Nicolas, *Publiees Sous les Auspices de l'Academie des Sciences*, Paris, Gauthier-Villars, 1898

Mal'tsev, Anatolii Ivanovich, *Metamathematics of Algebraic Systems; Collected Papers*: 1936-1967, Amsterdam, North-Holland, 1971

Napier, John, 1550-1617, *Napier's Mathematical Works* (Translated by William Francis Hawkins), Auckland, University of Auckland, 1982

Oka, Kiyoshi, *Oka Kiyoshi Sensei Ikoshu*, Kyoto-shi, Showado, 1980

Pappus, of Alexandria, *Pappi Alexandrini Mathematicae Collectiones a Federico Commandino Urbinate in Latinum Conversae, & Commentarijs Illustratae*, Bononiae, Duccijs, 1660

Quine, Willard Van Orman, *Selected Logic Papers*, New York, Random House, 1966

Ramanujan Aiyangar, Srinivasa, 887-1920, *Lost Notebook and other Unpublished Papers* (Introduction by George Andrews), Berlin, Springer, 1988

Sauri, Abbe, (Jean),1741-1785, *Cours Complet de Mathematiques*, Paris, Aux Depens de Ruau, 1744

Tarski, Alfred, Logic, *Semantics, Metamathematics: Papers*, Oxford, Clarendon Press, 1956

Ulam, Stanislaw M., *Sets, Numbers, and Universes: Selected Works*, Cambridge, Mass, MIT Press, 1974

Vaidyanathaswamy, R., *Collected Papers* (Foreword by V. S. Kirshnan, Madras, University of Madras, 1957

Waerden, Bartel Leendert van der, *Zur Algebraischen Geometrie*, Berlin, Springer-Verlag, 1983

Xiaotong, Wang, AD 580–640, *Jigu Suanjing* (王孝通), Edited and Translated B. Qian, *Ten Mathematical Classics*, Beijing, 1963

Yule, George Udny, 1871-1951, *Statistical Papers of George Udny Yule*, London, Griffin, 1971

Zeeman, E. C., *Catastrophe Theory: Selected Papers, 1972-1977*, Peoria, Addison-Wesley Pub., 1977

Index

A

D

E

F

G

H

I

S

T

[30] The Indexor (for want of a better term for a person employed to index a book) demanded a pay-raise upon having exhausted the original quantity of numbers originally assigned to him. Being that a book on advanced mathematics seldom gets on the New York Times Bestsellers List (or ever makes a killing selling movie production rights to Hollywood), I was forced to do without his continued services.

Z

* Mentioned Somewhere in a Footnote

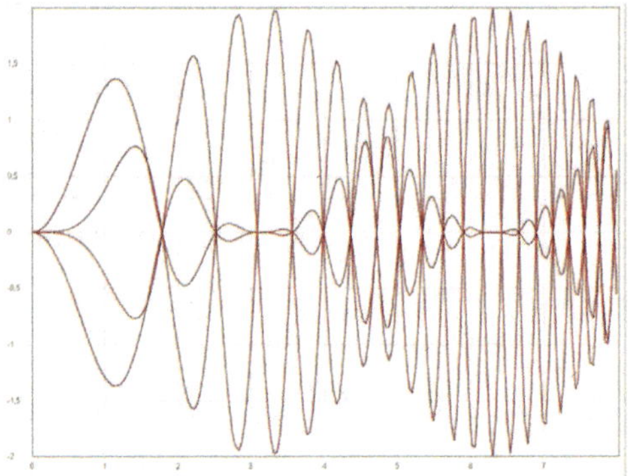

[Intended to be Left Blank, But Meaningless Graphic Put in Anyway

Made in the USA
Monee, IL
03 November 2026

67377426R00056